ENGINEERING TOOLS, TECHNIQUES AND TABLES

STRUCTURAL HEALTH MONITORING IN AUSTRALIA

ENGINEERING TOOLS, TECHNIQUES AND TABLES

Additional books in this series can be found on Nova's website under the Series tab.

Additional E-books in this series can be found on Nova's website under the E-books tab.

ENGINEERING TOOLS, TECHNIQUES AND TABLES

STRUCTURAL HEALTH MONITORING IN AUSTRALIA

TOMMY H.T. CHAN
AND
DAVID P. THAMBIRATNAM
EDITORS

Nova Science Publishers, Inc.

New York

NOTICE TO THE READER

The Publisher has taken reasonable care in the preparation of this book, but makes no expressed or implied warranty of any kind and assumes no responsibility for any errors or omissions. No liability is assumed for incidental or consequential damages in connection with or arising out of information contained in this book. The Publisher shall not be liable for any special, consequential, or exemplary damages resulting, in whole or in part, from the readers' use of, or reliance upon, this material. Any parts of this book based on government reports are so indicated and copyright is claimed for those parts to the extent applicable to compilations of such works.

Independent verification should be sought for any data, advice or recommendations contained in this book. In addition, no responsibility is assumed by the publisher for any injury and/or damage to persons or property arising from any methods, products, instructions, ideas or otherwise contained in this publication.

This publication is designed to provide accurate and authoritative information with regard to the subject matter covered herein. It is sold with the clear understanding that the Publisher is not engaged in rendering legal or any other professional services. If legal or any other expert assistance is required, the services of a competent person should be sought. FROM A DECLARATION OF PARTICIPANTS JOINTLY ADOPTED BY A COMMITTEE OF THE AMERICAN BAR ASSOCIATION AND A COMMITTEE OF PUBLISHERS.

Additional color graphics may be available in the e-book version of this book.

LIBRARY OF CONGRESS CATALOGING-IN-PUBLICATION DATA
Structural health monitoring in Australia / editors, Tommy Chan and David P. Thambiratnam.
p. cm.
Includes bibliographical references and index.
ISBN 978-1-61728-860-9 (hardcover)
1. Structural health monitoring--Australia. I. Chan, Tommy. II. Thambiratnam, David P.
TA656.6.S77 2010
624.1'71--dc22
2010026183

Published by Nova Science Publishers, Inc. † New York

CONTENTS

FOREWORD

STRUCTURAL HEALTH MONITORING: A ROAD BRIDGE OWNERS PERSPECTIVE

As we enter the beginning of the 21st Century, bridge owners worldwide are in an environment where monetary resources to maintain and/or replace infrastructure are limited and are approaching critical levels. External pressures to increase freight efficiency means legislators are requested to increase vehicle legal loads. It is critical that an asset owner is knowledgeable and is aware of the real condition of bridges. Recent bridge collapses at Montreal and Minnesota have highlighted the consequences of failure to manage road infrastructure.

Structural Health Monitoring (SHM) is critical for asset management to determine the actual structural capacity of the bridge, the deterioration rate and if the bridge is safe. It is considered prudent that asset owners of large iconic bridges have SHM on these structures to cost-effectively technically and financially manage these structures. Similarly, a selected sample of routine bridges should have SHM installed to ensure these bridges have long-term response similar to the design assumption as expected in the design.

SHM is a new emerging technology. The current technology includes:

- Optical sensors and fibre optic strain
- Wireless sensors
- Dynamic sensors
- Acoustic emission
- Vibration
- 3D laser scanning
- Micro electromechanical sensors (MEMS)

Where MEMS includes:

- Temperature
- Relative humidity
- Barometric pressure
- Light
- Sound

- Acceleration
- Pressure
- Inclination

Currently, the Queensland Department of Transport and Main Roads (DTMR) are planning on implementing SHM in:

- Fiber composite girders with optical sensors. These girders are new technology being developed for timber bridge replacement projects, and it is essential to fully understand their long-term response.
- Modular joint in the Gateway Bridge to include SHM to confirm design assumption and in-service behaviour.
- Routine deck unit bridges.

The development of the Australian National Structural Health Monitoring (ANSHM) Group is a major step in the right direction of developing Structural Health Monitoring in Australia. In my opinion, the future of SHM in Australia requires:

- Need for coordinated/co-operation development of SHM in Australia to harness energy and innovation.
- Close cooperation between universities, industry and asset owners.
- Specialist capability to monitor problem bridges.
- Professional advice on the extent of SHM that should be installed in routine DTMR bridges to monitor their health.

Dr. Ross Pritchard
Executive Director (Structures)
Department of Transport and Main Roads, Queensland

PREFACE

Structural Health Monitoring (SHM) is defined as the use of on-structure sensing system to monitor the performance of the structure and evaluate its health state. Recent bridge failures, such as the collapses of the 1-35W Highway Bridge in USA, the collapse of the Can Tho Bridge in Vietnam and the Xijiang River Bridge in Mainland China, all of which happened in the year 2007, have alerted to the importance of structural health monitoring. During the last two decades, SHM has attracted enormous research efforts around the world because it targets in monitoring structural conditions to prevent catastrophic failure and to provide quantitative data for engineers and infrastructure owners to design reliable and economical asset management plans.

This book gives a background of SHM technologies together with its latest developments and successful applications and compiles the research carried in these areas by some of the experts in Australian universities.

It is a book that is launched to celebrate the establishment of the Australian Network of Structural Health Monitoring (ANSHM). The network composing leading SHM experts in Australia aims to promote and advance SHM application, education, research and development in Australia. Although the title is Structural Health Monitoring in Australia, the technologies described can be applied worldwide.

Dr Ross Prichard, a practicing bridge engineer and road bridge owner, confirms in his forward for the book the importance of SHM. He also provides some inspiring insight for the future directions of SHM.

Structural health monitoring has been accepted as a justified effort for long-span bridges, which are critical to a region's economic viability. As the most heavily instrumented bridge project in the world, WASHMS— Wind And Structural Health Monitoring System—has been developed and installed on the cable-supported bridges in Hong Kong (Wong and Ni, 2009a). Chapter 1 aims to share some of the experience gained through the operations and studies on the application of WASHMS. It is concluded that SHM should be composed of two main components: Structural Performance Monitoring (SPM) and Structural Safety Evaluation (SSE). As an example to illustrate how the WASHMS could be used for structural performance monitoring, the layout of the sensory system installed on the Tsing Ma Bridge is briefly described. To demonstrate the two broad approaches of structural safety evaluation— Structural Health Assessment and Damage Detection—three examples in the application of SHM information are presented. These three examples can be considered as pioneer works for the research and development of the structural diagnosis and prognosis tools required by SHM for monitoring and evaluation applications.

Chapter 2 describes how vibration-based damage detection techniques could be used to successfully identify damage locations. Dynamic computer simulation techniques are used to develop and apply a multi-criteria procedure, incorporating changes in natural frequencies, modal flexibility and the modal strain energy, for damage localization in beam and plate structures under single- and multiple-damage scenarios. Numerically simulated modal data obtained through finite element analyses are used to develop algorithms based on changes of modal flexibility and modal strain energy before and after damage and used as the indices for assessment of the state of structural health. The proposed procedure is illustrated through its application to flexural members such as beams and plates under different damage scenarios, and the results confirm its feasibility for damage assessment.

Chapter 3 discusses the fact that bridges will deteriorate with age and often are subjected to additional loads or different load patterns than originally designed for. These changes in loads can cause localized distress and may result in bridge failure if not corrected in time. Visual inspection alone is not capable of locating and identifying all signs of damage, hence a variety of structural health monitoring (SHM) techniques is used regularly nowadays to monitor performance and to assess condition of bridges for early damage detection. Acoustic emission (AE) is one such technique that is finding an increasing use in SHM applications of bridges all around the world. This chapter presents the theory behind the acoustic emission technique, wave nature of AE, some previous applications and remaining challenges in its use as a SHM technique. Scope of the project currently undertaken and work carried out so far is also explained, followed by a brief description work to be completed and some recommendations for future work.

With support from the Australian Research Council (ARC), Cooperative Research Center for Integrated Engineering Asset Management (CIEAM) and Main Roads WA, intensive research has been carried out in the School of Civil and Resource Engineering, the University of Western Australia (UWA) on various aspects of structural condition monitoring. These include sensor development, signal-processing techniques, guided-wave (GW) propagation methods, vibration-based methods, model-updating methods, and integrated local GW and global vibration-based methods. The performance of these techniques and methods are all affected by unavoidable noises and uncertainties in structural modeling and response measurements. Because uncertainties may significantly affect the reliability of SHM, research efforts have also been spent on modeling and quantifying some uncertainties associated with SHM. Chapter 4 reports some of our research results related to Finite Element (FE) modeling errors, measurement noises and uncertainties associated with operational environments and signal-processing techniques and reliabilities of different damage indices for SHM. Methods for modeling these uncertainties in SHM are also presented and discussed. The results presented in this Chapter can be used to quantify possible uncertainties for better SHM.

Although non-destructive assessment techniques have been used for evaluating structural conditions of ageing infrastructure, such as timber bridges, utility poles and buildings, for the past 20 years, yet they face increasing challenges as the results of poor maintenance and inadequate funding. Replacement of structures such as an old bridge is neither viable nor sustainable in many circumstances. Hence, there is an urgent need to develop and utilize state-of-the-art techniques to assess and evaluate the "health state" of existing infrastructure and to be able to understand and quantify the effects of degradation in regard to public safety. Chapter 5 presents an overview of research work carried out by the authors in developing and implementing several vibration methods for evaluation of damage in timber bridges and utility poles. The technique of detecting damage involved the use of vibration methods, namely damage index method, which also incorporated artificial neural networks for timber bridges and time-based non-destructive evaluation (NDE) methods for timber utility poles. The projects involved successful numerical modeling and good experimental validation for the proposed vibration methods to detect damage in simple beams subjected to single and multiple damage scenarios and was then extended to a scaled timber bridge constructed and tested under laboratory conditions. The time-based NDE methods also showed promising trends for detecting the embedded depth and condition of timber utility poles in early stages of that research.

The Civionics Research Centre at the University of Western Sydney (UWS) is a multi-disciplinary research group with expertise in Civil/Structural Engineering, Electrical Engineering, Telecommunication Engineering, Construction Engineering and Mechanical Engineering. Multi-disciplinary approaches are being developed to advance the current capabilities of structural health monitoring. Chapter 6 briefly introduces some recent activities on research, development and implementation of structural health monitoring for civil infrastructure in the centre. The content includes wireless sensor network development, advanced signal processing and structural condition assessment.

Reliability-based safety assessment is used frequently in Europe and elsewhere to assess the need for maintenance and the remaining service life of structures. Chapter 7 describes another two important aspects in SHM, structural reliability analysis and service life prediction. In order to assess the effect of corrosion, quantitatively, an experimental study was conducted using an accelerated corrosion testing technique. Vibration tests were carried out fortnightly to study its effect on the natural frequency of RC beams subject to corrosion. One beam was taken out and broken every four weeks. The mass losses of steel rebar were measured to determine the corrosion state. The experimental results are used to develop an empirical model, which describes the relationship between natural frequency and corrosion loss. The statistics for model error for this relationship were then inferred. A spatial time-dependent structural reliability analysis is developed to update the deterioration process and evaluate the probability of structural failure based on vibration results. The performance of RC beams is used to illustrate the reliability analysis developed in this chapter. The inspection finding considered herein is the timing of test and the vibration results, which are used to provide an updated estimate of structural reliability. It was found that vibration test findings change the future reliability predictions significantly. A description of structural reliability and reliability-based safety assessment are also provided, where failure probabilities are compared with a typical target failure probability to illustrate how condition assessment findings can be used to more accurately assess, and often increase, service life predictions.

Bridge Management Systems (BMSs) have been developed since the early 1990s, as a decision support system (DSS) for effective Maintenance, Repair and Rehabilitation (MR&R) activities in a large bridge network. Historical condition ratings obtained from biennial bridge inspections are major resources for predicting future bridge deteriorations through BMSs. However, available historical condition ratings are very limited in all bridge agencies. This constitutes the major barrier for achieving reliable future structural performances. To alleviate this problem, a Backward Prediction Model (BPM) technique has been developed to help generate missing historical condition ratings. Its reliability has been verified using existing condition ratings obtained from the Maryland Department of Transportation, USA. This is achieved through establishing the correlation between known condition ratings and related non-bridge factors such as climate and environmental conditions, traffic volumes and population growth. Such correlations can then be used to determine the bridge condition ratings of the missing years. With the help of these generated datasets, the currently available bridge deterioration model can be utilized to more reliably forecast future bridge conditions. In this final chapter, the prediction accuracy based on 4 and 9 BPM-generated historical condition ratings as input data are also compared, using traditional bridge deterioration modeling techniques, i.e., deterministic and stochastic methods. The comparison outcomes indicate that the prediction error decreases as more historical condition ratings are available. This implies that the BPM can be utilized to generate unavailable historical data, which is crucial for bridge deterioration models to achieve more accurate prediction results. Nevertheless, there are considerable limitations in the existing bridge deterioration models, and further research is essential to improve their prediction accuracy.

It can be seen that this book, which covers most of the important aspects of SHM, will help students, engineers, researchers, bridge managers and bridge owners to have a better understanding on the hot topic of SHM. It can also be used as a textbook for courses on Structural Health Monitoring.

Tommy H.T. Chan & David P. Thambiratnam
Faculty of Built Environment and Engineering
Queensland University of Technology
Editors

Chapter 1

STRUCTURAL HEALTH MONITORING FOR LONG-SPAN BRIDGES – HONG KONG EXPERIENCE AND CONTINUING ONTO AUSTRALIA

Tommy H.T. Chan[*1], *K.Y. Wong*[*2], *Z.X. Li*[*3], *and Y.Q. Ni*[¥4]

[1]Queensland University of Technology, Queensland, Australia
[2]Highways Department, The Government of the Hong Kong Special Administrative Region, Tsing Yi, New Territories, Hong Kong
[3]College of Civil Engineering, Southeast University, Nanjing, PR China
[4]The Hong Kong Polytechnic University, Hung Hom, Kowloon, Hong Kong

ABSTRACT

Structural health monitoring has been accepted as a justified effort for long-span bridges, which are critical to a region's economic vitality. As the most heavily instrumented bridge project in the world, WASHMS—Wind And Structural Health Monitoring System—has been developed and installed on the cable-supported bridges in Hong Kong (Wong and Ni, 2009a). This chapter aims to share some of the experience gained through the operations and studies on the application of WASHMS. It is concluded that Structural Health Monitoring should be composed of two main components: Structural Performance Monitoring (SPM) and Structural Safety Evaluation (SSE). As an example to illustrate how the WASHMS could be used for structural performance monitoring, the layout of the sensory system installed on the Tsing Ma Bridge is briefly described. To demonstrate the two broad approaches of structural safety evaluation—Structural Health Assessment and Damage Detection—three examples in the application of SHM information are presented. These three examples can be considered

* School of Urban Development, Queensland University of Technology, GPO Box, 2434 Queensland 4001, Australia (tommy.chan@qut.edu.au)
‡ Bridges & Structures Division, Highways Department, The Government of the Hong Kong Special Administrative Region, The Administration Building, Northwest Tsing Yi Interchange, Tsing Yi, New Territories, Hong Kong (sebta.bstr@hyd.gov.hk)
≠ College of Civil Engineering, Southeast University, Nanjing 210018, PR China (zhxli@seu.edu.cn)
¥ Department of Civil and Structural Engineering, The Hong Kong Polytechnic University, Hung Hom, Kowloon, Hong Kong (Yiqing.Ni@polyu.edu.hk)

as pioneer works for the research and development of the structural diagnosis and prognosis tools required by the structural health monitoring for monitoring and evaluation applications.

INTRODUCTION

Structural health monitoring has been accepted as a justified effort for long-span bridges, which are critical to a region's economic vitality. For example, Skarnsundet Bridge (Myrvoll et al., 1996) in Norway was instrumented with a fully automatic data acquisition system to monitor wind, acceleration, inclination, strain, temperature and dynamic displacements of the bridge during construction and in service. An online monitoring system consisting of 65 sensors and a signal processor has been developed for New HaengJu Bridge (Chang and Kim, 1996) in Korea. And a fiber optical surveillance system consisting of 14 optical sensors and 26 electrical sensors was installed on Storck's Bridge in Switzerland (Nellen et al., 1997). As the most heavily instrumented bridge project in the world, WASHMS—Wind And Structural Health Monitoring System—has been developed and installed on the cable-supported bridges in Hong Kong (Wong and Ni, 2009a). This chapter aims to share some of the experience gained through the operations and studies on the application of WASHMS.

WASHMS IN HONG KONG

WASHMS is the acronym of the Wind And Structural Health Monitoring System, which is a structural health monitoring system installed at each of the six cable-supported bridges in Hong Kong. These bridges include the Tsing Ma (suspension) Bridge, the Kap Shui Mun (cable-stayed) Bridge, the Ting Kau (cable-stayed) Bridge, the two cable-stayed bridges in Hong Kong—Shenzhen Western Corridor, and the Stonecutters (cable-stayed) Bridge.

The WASHMS has been deployed and periodically updated for effectively executing the functions of structural condition monitoring and structural degradation evaluation as well as for the safe and economic operation of these structures. To summarize, the objectives of the WASHMS are:

i.   to have a better understanding of the structural behavior of the cable-supported bridges under their in-service conditions;

ii. to develop, validate and harden the bridge evaluation techniques based on measurement results;

iii. to setup/calibrate/update the monitoring and evaluation models and criteria for describing and limiting the variation ranges of the physical parameters that influence bridge performance under in-service conditions;

iv. to evaluate structural integrity immediately after rare events such as typhoons, strong earthquakes and vehicle/vessel collisions, etc.;

v. to provide information and analytical tools for the planning, scheduling, evaluating and designing of effective long-term bridge inspection and maintenance strategies, hence promoting the bridge maintenance activities from corrective status to preventive status; and

vi. to minimize the lane-closure time required for exercising bridge inspection and maintenance activities, hence maximizing the traffic operational period of the bridges.

It can be seen that in order to achieve the above-mentioned objectives, WAHSMS is designed to comprise a lot of sensors. As an illustration, Figure 1 shows the instrumentation layout of the Tsing Ma Bridge (TMB), which includes the layout of the sensory systems and their associated data acquisition outstations. Its monitoring categories could be divided into three areas, namely: environment, traffic loads, bridge responses (including bridge features). Tables 1-3 show, respectively, the corresponding sensory systems and monitoring physical parameters in each of these categories. Wong and Ni (2009b) give a very detailed explanation of how the monitoring results obtained from WAHSMS could be used to assess the health status of TMB using structural diagnosis approaches. For the instrumentation layout of the other five cable-supported bridges in Hong Kong, please refer to Wong and Ni (2009a).

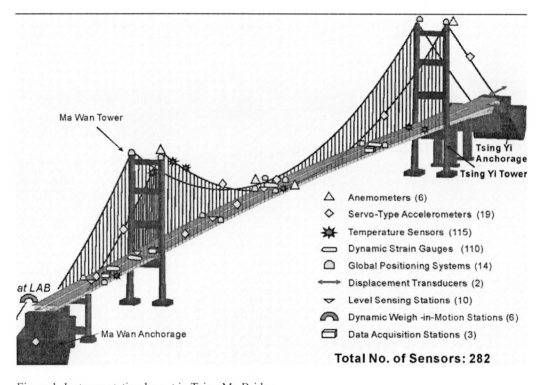

Figure 1. Instrumentation layout in Tsing Ma Bridge

Table 1. Sensory Systems and Physical Parameters for Processing and Derivation (Environmental Category)

Physical Quantity	Required Types of Sensory Systems	Physical Parameters for Processing and Derivation
Wind Loads Monitoring	• Ultrasonic-Type Anemometers (usually at deck levels) • Propeller-Type Anemometers (usually at tower-tops)	• Wind Speeds & Wind Direction Plots (Time-Series Data) • Wind Speeds (Mean & Gust) & Directions (Histograms) • Terrain Factors and Wind Speed Profile Plots • Wind Rose Diagrams • Wind Incidences at Deck Level • Wind Turbulence Intensities and Intensity Profile Plots • Wind Turbulent Time and Length Scale Plots • Wind Turbulent Spectrum and Co-Spectrum Plots • Wind Turbulent Horizontal & Vertical Coherence Plots • Wind Response and Wind Load Transfer Function • Wind Induced Accumulated Fatigue Damage • Histograms of Air Pressure, Rainfall and Humidity
Temperature Loads Monitoring	• Platinum RTD Type for Air, Concrete, Structural Steel, Asphalt Pavement • Thermo-Couplers for Main Cable	• Effective Temperatures in Tower, Deck & Cables • Differential Temperatures in Deck & Tower • Air Temperatures and Asphalt Pavement Temperatures • Temperature Response and its Temperature Load Transfer Function
Seismic Loads Monitoring	• Fixed Servo-Type Accelerometers	• Acceleration Spectra near Towers & Anchorages • Deck and Tower Response Spectra • Seismic Response and its Seismic Load Transfer Function

Table 2. Sensory Systems and Physical Parameters for Processing and Derivation (Traffic Load Category)

Physical Quantity	Required Types of Sensory Systems	Physical Parameters for Processing and Derivation
Highway Traffic Loads Monitoring	• Dynamic Weigh-in-Motion Stations (Bending-Plate Type) • Dynamic Strain Gauges • CCTV Cameras	• Gross Vehicular Weight Spectrum per Traffic Lane • Axle Weight (AW) Spectrum per Traffic Lane • Equivalent No. of Standard Fatigue Vehicles (SFV) Spectrum per Traffic Lane • Equivalent No. of Standard Fatigue Axle (SFA) Spectrum per Traffic Lane • Highway Induced Accumulated Fatigue Damage -SFV • Highway Induced Accumulated Fatigue Damage - SFA • Over-Load Vehicles Detection • Traffic Composition per Traffic Lane • Traffic Load Response, and its Traffic Load Transfer Function
Railway Traffic Loads Monitoring	• Dynamic Strain Gauges • CCTV Video Cameras	• Bogie Loads per Line of Train • Train Loading Spectrum • Equivalent Standard Load (Train) Spectrum • Train Induced Accumulated Fatigue Damage • Train Load Response, and its Train Load Transfer Function

**Table 3. Sensory Systems and Physical Parameters for Processing and Derivation
(Bridge Responses including Bridge Features Category)**

Physical Quantity	Required Types of Sensory Systems	Physical Parameters for Processing and Derivation
Static Influence Coefficients Monitoring	• Level Sensing Stations • GPS • Dynamic Strain Gauges	• Stress History of each Vehicular Type per Traffic Lane • Stress Range of each Vehicular Type per Traffic Lane • Influence Surfaces for Combined Deck Plates and Troughs • Stress History of each Type of Train per Rail Line • Stress Range of each Type of Train per Rail Line • Influence Coefficients at Tower-Tops & Deck Mid-Span
Dynamic Features Monitoring	• Fixed & Portable Servo-Type Accelerometers	• Global Bridge Modal Frequencies • Global Bridge Mode Shapes • Global Bridge Modal Damping Ratios (Derived) • Global Bridge Modal Mass Participation Factors (Derived) • Cable Frequencies
Cable Forces Monitoring	• Portable Servo-Type Accelerometers	• Cable Frequencies and hence Cable Forces • Cable Damping Ratios
Geometry Monitoring	• GPS • Level Sensing Stations • Displacement transducers • Servo-Type Accelerometers • Static Strain gauges	• Thermal Movements of Cables, Deck & Towers • Wind Movements in Cables, Deck & Towers • Seismic Movements in Deck & Towers • Highway Load Movement in Deck & Cables • Railway Load Movement in Deck & Cables • Creep & Shrinkage Effects in Concrete Towers
Stress Monitoring	• Dynamic Strain Gauges • Static Strain gauges	• Stress Historical Plots of Instrumented Components • Stress Demand Ratio Plots of Instrumented Components • Principal Stresses Plots of Instrumented Components • Force Demand Ratios Plots of Concrete-Steel Interfaces
Fatigue Life Monitoring	• Dynamic Strain Gauges	• Total Accumulated Fatigue Damage; hence Remaining Fatigue Life due to Combined Load-Effects • Accumulated Fatigue Damage, hence Remaining Fatigue Life Estimation due to Individual Load-Effects in Structural Components
Articulation Monitoring	• Dynamic Strain Gauges • Displacement Transducers	• Stress Histories in Bearings • Stress Demand Ratios in Bearing • Motion Histories in Movement Joints

RESEARCH WORK ON STRUCTURAL HEALTH MONITORING

Structural health monitoring has received a great deal of attention from many researchers over the last thirty years, as evidenced by the large amount of published literature on the topic summarized in comprehensive literature reviews by Doebling et al. (1996), the Hong Kong Polytechnic University (1998) and Sohn et al. (2004).

Most often, researchers relate Structural Health Monitoring with damage detection that could be reflected by how they define Structural Health Monitoring, e.g.,

> Structural health monitoring is "the use of in-situ, nondestructive sensing and analysis of structural characteristics, including the structural response, for the purpose of detecting changes that may indicate damage or degradation" (Housner et al., 1997).

> Structural health monitoring is the "process of implementing a damage identification strategy for aerospace, civil and mechanical engineering infrastructure" (Farrar and Worden, 2007; Sohn et al., 2004).

> "The process of implementing a damage identification strategy for aerospace, civil and mechanical engineering infrastructure is referred to as structural health monitoring (SHM)" (Worden et al., 2007)

> Structural health monitoring "can be defined as tracking the responses of a structure along with inputs, if possible, over a sufficiently long duration to determine anomalies, to detect deterioration and to identify damage for decision making" (Catbas et al., 2008).

However, from our experience in Hong Kong, SHM should not be confined to damage detection only. Sometimes, one could even find that existing damage detection approaches will not be effective in the SHM of large complicated civil structures. Actually, Structural Health Monitoring should be defined as the use of on-structure sensing system to monitor the performance of the structure and evaluate its health state. Hence, Structural Health Monitoring should be composed of two components: Structural Performance Monitoring (SPM) and Structural Safety Evaluation (SSE). Structural Performance Monitoring refers to the monitoring (observation) of structural performance in structure and its components under its (their) designated performance limits (or criteria) at serviceability limit states (SLS) by on-structure instrumentation system; whereas Structural Safety Evaluation refers to the evaluation of possible damage in structure or its components and/or the assessment of its health status by analytical tools, which are developed and calibrated in the course of structural health monitoring, basing on its (their) designated performance limits at ultimate limit states (ULS).

The structural performance limits at SLS are adopted as the basis of SPM, which is an important component of Structural Health Monitoring. Even it could be standalone in an effective Structural Health Monitoring system. This is because the serviceability load combinations specified in most limit state design codes for bridgework are reasonable estimates of combinations of loads likely to cause maintenance problem or affect the performance of the structure and its components, and the probability of exceeding the serviceability load combination is approximately 63.36% in 120-year return period basing on 120-year design life (and the value will be roughly the same for a 100-year design life in the

Australian context). The value of probability of exceedance is normally considered as the standard for loads and load-effects determination at SLS for most bridge design codes and standards in the world.

The performance of the instrumentation system for structural health monitoring should be devised in accordance with the structural performance limits at SLS, and the analytical tools for structural safety evaluation should be developed and calibrated to identify and quantify the existence and cause of damage respectively, based on the structural performance limits defined at ULS. Such approach makes the structural health monitoring to be accepted by bridge engineers because its design and operation are complied with the codified requirements for bridge design, construction and maintenance—the legal basis for monitoring and evaluation.

Hence, other than structural safety evaluation, the key functions of development and deployment of a Structural Health Monitoring, are (i) to improve/enhance the current practice of bridge inspection and maintenance from local and subjective condition to global and objective condition; and (ii) to provide data and information for updating/amending the contemporary bridge design manuals, standards and codes which are the standards for future works of : (a) design & construction of new bridges and (b) maintenance & rehabilitation of old bridges. It can be seen that the objectives of WASHMS described earlier aligned neatly with these functions. Hence, all methods of Structural Health Monitoring must be complied with the codified requirements otherwise the bridge engineers will have much reluctance for its acceptance in their bridge inspection and maintenance works.

Research on Structural Health Monitoring should advance the technologies and enhance the development on both SPM and SSE. Most of the past and recent research works on Structural Health Monitoring have been concentrated mostly on damage detection. It is timely that we need to reconsider our directions to carry out research on SHM. Described below are three of our previous research studies in SHM, which cover both the structural performance monitoring and structural safety evaluation:

i. Damage Detection using Neural Network (Damage Detection under SSE)
ii. Health Assessment of Steel Bridges (SPM and Health Assessment under SSE)
iii. Use of Multi-scale modeling in Fatigue Assessment (SPM and Health Assessment under SSE)

DAMAGE DETECTION USING NEURAL NETWORK

As mentioned above, the damage detection approach is one of the two broad approaches used in SHM. The difficulties encountered in the damage assessment of complex civil engineering structures using vibration measurements are primarily due to measurement noise, modeling error, uncertainty of ambient conditions, insensitivity of modal properties of redundant structures to local damage, and incompleteness of measured data. The majority of vibration-based damage identification methods require a refined or validated analytical or finite element model as the baseline reference (Ko and Ni, 1999). The *refined* or *validated* model means that the measured and analytical modal properties of the undamaged structure are in agreement or perfectly correlated. Consequently, the damage incurred in any structural component can be directly simulated in the model. Measurement noise and structural uncertainty are also inevitable in addressing the damage detection of civil structures. The sources of structural uncertainty includes uncertainties inherent in the materials and construction of civil structures, the variability of structural properties due to environmental conditions (temperature and humidity), and the variability of mass loading present in civil structures.

A robust damage identification method should have a considerable tolerance to measurement noise, modeling error and variability of structural properties due to ambient conditions. In the following, the applicability of the auto-associative neural network for damage detection of the three earlier constructed cable-supported bridges—Tsing Ma Bridge, Kap Shui Mun Bridge and Ting Kau Bridge—using noise-corrupted modal data is examined through numerical simulations. An auto-associative network is designed as a novelty filter to detect the occurrence of damage. The damage identification algorithm in terms of the auto-associative neural network has an advantage that they can accommodate to the noise-corrupted data and the series data in a convenient way.

Neural Network Based Novelty Detection

The object of novelty detection is to establish if a new pattern differs from previously obtained patterns in some significant respect. This can be realized by using an auto-associative memory neural network. An auto-associative network can be a multi-layer feed-forward perceptron with "bottleneck" hidden layer(s) (Chan et al., 2000; Worden, 1997). This network is trained to reproduce at the output layer, the patterns that are presented at the input layer. Thus, the output layer must have the same number of the input nodes. However, the input values will not be perfectly reconstructed in the output. Since the patterns are passed through hidden layers that have fewer nodes than the input layer, the network is forced to learn just the significant prevailing features of the patterns. The central hidden layer, which acts as a bottleneck, generates an internal representation that compresses redundancies in the input pattern while retaining important information to the output.

When the auto-associative network is used for anomaly detection or damage alarming, a series of measurement data of the healthy structure under normal conditions are required as both input and output to train the network. No information on the structural model is needed. After the network is trained, the input data presented on training are passed again into the trained network to yield a set of output data. The difference between the input and output vectors is then measured using some form of distance function, called *novelty index*. In the testing phase, a new series of measurement data obtained later from the same structure (damaged or undamaged) are passed into the above network to form a novelty index sequence of testing phase. If this sequence deviates from the novelty index sequence of training phase, the occurrence of damage is alarmed.

In the present study, an improved novelty index is formulated for damage alarming. In the training phase, a series of measured modal data of the healthy (intact) structure, $f = \{f_1\ f_2 \cdots\cdots f_N\}^T$, are used as input to the network. The output $y = \{y_1\ y_2 \cdots\cdots y_N\}^T$ of the network is set as follows

$$y_i = (f_i - m_i)\alpha + m_i, \qquad i = 1, 2, \cdots, N \tag{1}$$

where α is a positive constant and is taken as three in the present study, m_i is the mean of the ith element f_i of the input vector f over the training data. It is worth noting that training the network needs just the measurement data of the healthy structure. After performing the training, the input pattern f presented on training is fed again into the trained network to yield output pattern $\hat{y}$, and the novelty index sequence for the training phase, $\lambda(y)$, is obtained in terms of the Euclidean distance as

$$\lambda(\mathbf{y}) = \|\hat{\mathbf{y}} - \mathbf{y}\| \tag{2}$$

In the testing phase, a new series of modal data, $f_t = \{f_{1t} f_{2t} \cdots\cdots f_{Nt}\}^T$, are measured from the same structure (damaged or undamaged), and are passed into the above trained network to yield output $\hat{y}_t$. The corresponding novelty index for the testing set is then obtained by

$$\lambda(\mathbf{y}_t) = \|\hat{\mathbf{y}}_t - \mathbf{y}_t\| \tag{3}$$

where $\mathbf{y}_t = \{y_{1t} y_{2t} \cdots\cdots y_{Nt}\}^T$ is the vector with its ith element

$$y_{it} = (f_{it} - m_i)\alpha + m_i, \qquad i = 1, 2, \cdots, N \tag{4}$$

If the testing novelty index sequence deviates from the training novelty index sequence, the occurrence of damage is flagged; if the two sequences are indistinguishable, no damage is signaled. In order to quantitatively judge if damage occurs, a threshold δ_λ is introduced herein. It is estimated from the training data and is defined as

$$\delta_\lambda = \overline{\lambda} + 4\sigma_\lambda \tag{5}$$

where $\overline{\lambda}$ and σ_λ are the mean and standard deviation of the testing novelty index sequence over the training data respectively.

Damage Occurrence Detection of the Three Bridges

Three three-dimensional finite element models have been developed separately for the purpose of modal sensitivity analysis and damage detection simulation of the three bridges. Each of the developed model has the following attributes (HK PolyU, 1999 and Ko et al., 2002): (1) the model is accurate enough through comparison of the computed and measured modal parameters; (2) the spatial configuration of the original structure remains in this three-dimensional model; (3) the stiffness contribution of all individual structural components is independently described in the model, so the modal sensitivity to any structural component can be evaluated conveniently and accurately; (4) the towers and piers are modeled as Timoshenko's beam elements in which the shear and torsion deformations are considered; (5) the effect of the tension force in the cables on the stiffness is taken into account in the model through a nonlinear static iteration analysis. Table 4 shows the total number of nodes and elements involved in each of the three bridge models.

Table 4. Size of the 3-D Finite Element Models Developed

	No. of Nodes	No. of Elements
Tsing Ma Bridge (TMB)	7375	17677
Kap Shui Mun Bridge (KSMB)	6006	8581
Ting Kau Bridge (TKB)	4981	7661

Modal analysis shows that (HK PolyU, 1999) the vibration modes of the Tsing Ma Bridge can be classified into the following seven categories: (1) global vertical bending modes; (2) global lateral bending and torsion modes; (3) central span cable local sway modes; (4) Ma Wan side span cable local sway modes; (5) Tsing Yi side span free cable local modes (in-plane and out-of-plane); (6) Ma Wan side span deck dominated modes; and (7) tower dominated modes (sway, bending and torsion). This modal classification gives guidance to construct the auto-associative neural networks for anomaly detection. After merging categories (4) and (5) into one group and merging categories (6) and (7) into one group, the following five auto-associative networks are generated for the damage occurrence detection of the Tsing Ma Bridge: the first network is based on the natural frequencies of the global vertical bending modes [category (1)]; the second network is based on the natural frequencies of the global lateral bending and torsion modes [category (2)]; the third network is based on the natural frequencies of the central span cable local sway modes [category (3)]; the fourth network is based on the natural frequencies of the local modes of Ma Wan side span cables and Tsing Yi side span free cables [categories (4) and (5)]; and the fifth network is based on the natural frequencies of tower dominated modes and Ma Wan side span deck dominated modes [categories (6) and (7)].

Similarly, four networks are generated in the numerical simulation of damage occurrence detection on the Kap Shui Mun Bridge: the first network is based on the natural frequencies of the deck vertical bending modes [category (1) of KSMB]; the second network is based on the natural frequencies of the deck lateral and torsion modes [category (2) of KSMB]; the third network is based on the natural frequencies of the tower-dominated modes [category (3) of KSMB]; the fourth network is based on the natural frequencies of the cable local modes and miscellaneous modes [category (4) of KSMB]. Five networks are generated in the numerical simulation of damage occurrence detection on the Ting Kau Bridge: the first network is based on the natural frequencies of the global vertical bending modes [category (1) of TKB]; the second network is based on the natural frequencies of the global lateral bending modes [category (2) of TKB]; the third network is based on the natural frequencies of the global torsional modes [category (3) of TKB]; the fourth network is based on the natural frequencies of the cable local out-of-plane modes [category (4) of TKB]; the fifth network is based on the natural frequencies of cable local in-plane modes [category (5) of TKB].

A detailed modal correlation and modal sensitivity analysis have been made for the three bridges with respect to various assumed damage cases (e.g., for TMB as discussed in H.K.

PolyU, 1999; Wang et al., 2000). The modal influence patterns and sensitivity levels of different types of damage have been identified. Therefore, a comprehensive use of the above-fourteen networks in terms of different modal categories is conducive to both the detection of damage occurrence and the identification of damage type. All the fourteen auto-associative neural networks are four-layer feed-forward perceptrons with "bottleneck" internal layers. The activation functions are taken as the tan-sigmoid function between the second layer and the third layer, and the linear transfer function between the input layer and the second layer, and between the third layer and the output layer. For the Tsing Ma Bridge, the first network has a node structure 11-6-6-11, with the input set being the natural frequencies of the first 11 global vertical bending modes of the bridge (i.e., the 2nd, 3rd, 6th, 12th, 14th, 17th, 28th, 33rd, 39th, 48th, and 60th modes); the second network also has a node structure 11-6-6-11, with the input set being the natural frequencies of the first 11 global lateral bending and torsional modes of the bridge; the third network has a node structure 17-9-9-17, with the input set being the natural frequencies of the first 17 local sway modes of central span cables; the fourth network has a node structure 12-6-6-12, with the input set being the natural frequencies of the first four local sway modes of Ma Wan side span cables and the first eight local sway modes of Tsing Yi side span free cables; the fifth network has a node structure 9-5-5-9, with the input set being the natural frequencies of the first three Ma Wan side span deck dominated modes and the first six tower dominated modes. The node structures for the other two bridges are given in Table 5.

Table 5. Node Structures for the Kap Shui Mun Bridge and Ting Kau Bridge

	Node Structure				
	1st Network	2nd Network	3rd Network	4th Network	5th Network
Kap Shui Mun Bridge	14-11-11-14	12-9-9-12	7-5-5-7	12-9-9-12	N.A.
Ting Kau Bridge	11-6-6-11	6-4-4-6	12-9-9-12	18-11-11-18	13-9-9-13

For the simulation study, the analytical natural frequencies of the intact structures are obtained through modal analysis of the validated finite element models. Each frequency to be used in the networks is added by an independent normally distributed random sequence with mean and 0.005 variance to form a series of noisy/uncertain "measured" data of the healthy structure under normal conditions. This implies that the normal frequency fluctuation is about ±1.5% maximum error. Such generated frequency sequences, each with the length of 500 data, are regarded as noise-corrupted measurement data to train the fourteen auto-associative networks for the damage occurrence detection. The output sets to the networks are given by Eq. (1), and the networks are trained using the back-propagation algorithm. After performing the training, the input sets presented on training are fed again into the trained networks to yield the output, and the novelty index sequences for the training phase are obtained by Eq. (2). The fluctuation in these sequences reflects the measurement noise and uncertainty of ambient environments under normal conditions of the healthy bridge. The threshold values of the training novelty index sequences are computed with Eq. (5) and given in Table 6.

For the Tsing Ma Bridge, a total of 15 damage cases are introduced in the simulation testing. The first three cases simulate the damage to the lateral and horizontal bearings between the bridge tower and deck; the fourth and fifth cases are the damage to a side span cable and to an anchorage, respectively; the sixth to eighth cases simulate the damage to tower saddles and to tower cross-beams; the ninth to tenth cases are the damage to hangers; the eleventh to fourteenth simulate the damage at deck members; the fifteenth case simulates the damage at rail waybeams.

Table 6. Threshold δ_λ of Training Novelty Index Sequences Defined by Equation (5)

	δ_I	δ_{II}	δ_{III}	δ_{IV}	δ_V
Tsing Ma Bridge	0.01443	0.01269	0.01745	0.01737	0.01561
Kap Shui Mun Bridge	0.01857	0.01470	0.01301	0.02031	N.A.
Ting Kau Bridge	0.01853	0.01545	0.02321	0.02189	0.01929

By introducing each damage case into the finite element model, the modal properties of the damaged structure are analyzed. A series of "measurement" data of the damaged structure are then generated by adding normally distributed random sequences of 0.005 variance to the computed natural frequencies. They are passed into the above networks to achieve the novelty index sequences of testing phase. The novelty index sequences in the testing phase are obtained using Eqs. (3) and (4). If these sequences deviate from the novelty index sequences of training phase, the occurrence of damage is alarmed. As an example, Figure 2 shows the evaluated novelty indexes of Case 5 in terms of the five auto-associative networks. The novelty index sequences in the testing phase are depicted in the figure corresponding to the latter 200 data. In this case, the damage is simulated by 0.5m horizontal shift of one anchorage. It is seen that for indexes I, III and IV, the novelty index sequences in the testing phase and in the training phase are just distinguishable, but most of the novelty index values are less than the corresponding thresholds. For index II, which uses the frequencies of the global lateral bending and torsional modes, the novelty index sequence in the testing phase deviates significantly from the sequence in the training phase, unambiguously signaling the occurrence of damage. For index V, the damage occurrence cannot be flagged because the novelty index sequences in the testing phase and in the training phase are indistinguishable. It is found that in the noise level of 0.005 variance ($\pm 1.5\%$ maximum error), the novelty indexes cannot indicate the occurrence of damage for some damage cases. For Case 1 (damage at a vertical bearing), Cases 9 and 10 (damage at hangers) and Case 15 (damage at rail waybeams), the damage cannot be flagged by the novelty indexes. For the damage at deck members (Cases 11 to 14), the novelty indexes cannot detect the occurrence of the damage respectively at two bottom chords (Case 11), at two top and two bottom chords (Case 12), and at two diagonal members (Case 13). Only the damage that occurs simultaneously at two top chords, two bottom chords and two diagonal members (Case 14) can be detected where the novelty index sequences in the testing phase and in the training phase are just distinguishable. For Case 2 and 3 (damage at side-support bearings), Case 7 (damage at two tower saddles) and Case 8 (damage at a tower cross-beam), the novelty indexes clearly indicate the damage occurrence. For Case 4 (damage at one side span cable), the novelty index sequences in the testing phase and in the training phase are just distinguishable. It is interesting to relate the detectability by means of novelty indexes with the level of frequency change ratios caused by damage. Table 7 lists the evaluated maximum frequency change ratios of the first 60 modes of the bridge corresponding to the 15 damage cases. It is found that when the maximum frequency change ratio γ is less than 0.3% (Cases 1, 9, 10, 11, 12, 13 and 15), the damage occurrence cannot be flagged by the developed novelty indexes. This is because the maximum frequency change ratios in these damage cases are far less than the normal noise level (±1.5% maximum error). When the maximum frequency change ratio γ is between 0.3% and 1.0% (Cases 4 and 14), the damage is just detectable with a weak alarming signature. The novelty indexes can unambiguously alarm the damage states when the maximum frequency change ratio γ is greater than 1.0% (Cases 2, 3, 5, 6, 7, and 8).

Table 7. Damage Caused Maximum Frequency Change Ratio

Damage Case No.	Case 1	Case 2	Case 3	Case 4	Case 5
γ(%)	0.0873	2.3773	3.0048	0.4556	1.4968
Damage Case No.	Case 6	Case 7	Case 8	Case 9	Case 10
γ(%)	3.5823	7.2066	1.8622	0.0786	0.0672
Damage Case No.	Case 11	Case 12	Case 13	Case 14	Case 15
γ(%)	0.1688	0.2030	0.0693	0.8843	0.2037

Similar observations are found in the simulation studies of the Kap Shui Mun Bridge and Ting Kau Bridge.

USE OF SHM DATA FOR FATIGUE EVALUATION OF BRIDGES – STRUCTURAL HEALTH ASSESSMENT APPROACH

SHM data collected could be used for damage detection. However, SHM data could also be used for assessing the structural health conditions of a bridge to evaluate its structural safety. It would be an efficient way for assessing the health conditions by evaluating accumulative fatigue in bridges and to develop an approach of fatigue analysis by taking full advantage of the real-time monitoring data. Little information, however, exists in the literature about fatigue analysis using structural health monitoring data, although the practical implementation of the health monitoring system has been reported for several large cable-stayed bridges. Therefore, it is necessary to seek an efficient approach to accurately assess fatigue of these bridges by taking full advantage of online monitoring data measured by the system.

In this section, an effective approach for fatigue evaluation and detection of bridges with online structural health monitoring system will be studied. In order to develop a methodology and strategy for fatigue evaluation and detection of bridge decks with the structural health monitoring data, the strain-time history recorded by the structural health monitoring system for the studied bridge deck of the Tsing Ma Bridge at actual traffic loading is investigated. The effective stress range is used to evaluate the fatigue strength at different locations of the bridge deck and to detect the critical location for possible fatigue failure. Calculated results of the fatigue evaluation and detection of critical location are given for the deck section of the Tsing Ma Bridge.

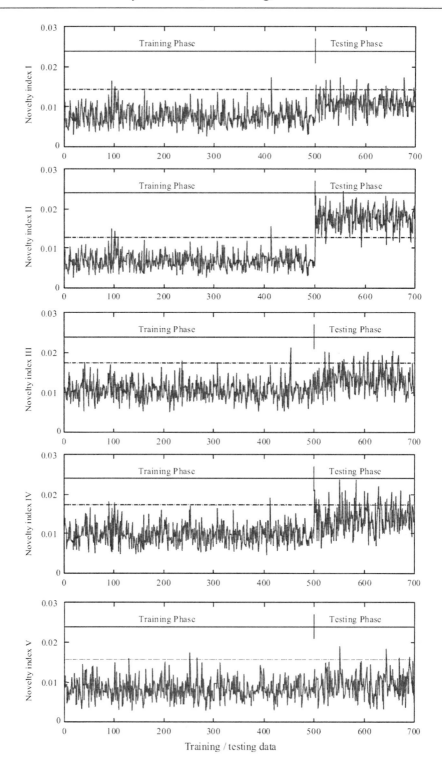

Figure 2. Novelty Indexes Evaluated on Testing Data of TMB Damage Case 5

Strategy for Fatigue Evaluation of Bridges with Structural Health Monitoring Data

Highway bridges are subjected to a large number of repetitive loads of different magnitudes that are caused primarily by the passage of vehicles. In most types of short- and medium-span bridges, each vehicle, especially a truck, produces a major stress cycle with superimposed vibration stresses that are much smaller than the major cycle. In long-span bridges, individual vehicles produce only very small stress cycles in the main members, but larger cycles may be produced when the entire bridge is subjected to lane loading during peak traffic. Thus, bridges are subjected to variable-amplitude stress cycles to be generally considered as occurred in a random sequence. However, it would be difficult to evaluate fatigue of bridges if the variable-amplitude stress cycles are directly used as a random sequence. Some necessary work on the pattern of stress or strain history should be done for the purpose of fatigue evaluation for the bridge with online measured data. The strain-time history data recorded at different locations and times, especially during peak traffic, are investigated here for the Tsing Ma Bridge.

As shown in Figure 3, the locations of strain gauges installed in the Tsing Ma Bridge include rail track sections at CH 24662.50, bridge-deck trough section at CH 24664.75 and deck at tower and rocker bearing links at CH 23623.00. The most critical parts of the cross-frames for fatigue damage have been identified during the design of the WASHMS.

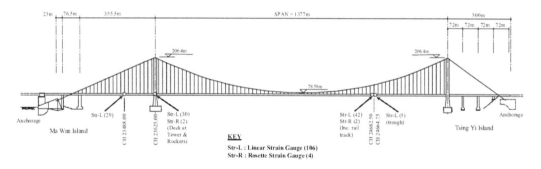

Figure 3. Strain Gauges Layout of the Tsing Ma Bridge

Strain data have been recorded at locations of strain gauges since the Tsing Ma Bridge was commissioned. It is a useful database for online fatigue assessment of the bridge. The measured data of strain-time history is observed to be very complicated, since the loading on the bridge is considered to be random. In order to find out the feature of the strain-time history of the Tsing Ma Bridge, investigations on the pattern of strain-time history measured by strain gauges are carried out for several days by recording continuously the strain 24 hours a day at a location, and that at different locations of the deck section. The strain history showed some common feature of strain or stress spectrum for bridges. A single major strain cycle produced by the passage of a vehicle over the bridge is superimposed with the vibration strain. The vibration strain may continue after the vehicle has left the bridge. It is much smaller than the major cycle and can be neglected in fatigue evaluation. Furthermore, following observation can be obtained from the investigation on the strain-time history of the Tsing Ma Bridge:

- Daily strain-time curves are similar from day to day. They have some common characteristics in curve shape and magnitude of cycles; for example, there are almost no strain pulses from 01:30 to 05:30 since rare vehicles are on the road of the bridge.
- Strain time curve in one hour can be considered to be composed of many small pulse of strain and some of higher pulse at the rate of approximate one high pulse per two minutes.
- Each pulse in strain time curve will be a strain cycle to generate cumulative fatigue damage. A lot of small cycles in strain time curve should be neglected in the fatigue calculation since the magnitude of related stress cycles must be less than the threshold value to effect fatigue.

The above feature of the strain history on the bridge deck of TMB suggests that the strain history of such bridges under traffic load can be approximately represented by a block of repeated cycles in which the cycles are daily repeated. The stress spectrum of the block, defined as the representative block of cycles, can be obtained by rain-flow counting cycles of the strain history and statistical analysis on daily samples of strain spectrum; so that the approach of fatigue analysis for deck-sections of bridges with the structural health monitoring system is provided as shown in Figure 4.

Detection of Critical Locations on Basis of Effective Stress Range

Extensive test results showed that variable-amplitude random sequence stress spectrums, such as those that occur in actual bridges, can be conveniently represented by a single constant-amplitude effective stress range that would result in the same fatigue life as the variable-amplitude stress range spectrum. The effective stress range concept can be used directly in the design of critical bridge members or in estimating the remaining fatigue life of existing bridges, and could eventually be incorporated in bridge-design specification (AASHTO, 1989, 1990).

The effective stress range for a variable-amplitude spectrum is defined as the constant-amplitude stress range that would result in the same fatigue life as the variable-amplitude spectrum. It can be written as:

$$\Delta\sigma_{ef} = \left[\frac{1}{N_T} \sum_i n_i \Delta\sigma_i^m \right]^{\frac{1}{m}} \tag{6}$$

in which, n_i = number of cycles of stress range $\Delta\sigma_i$;

$\Delta\sigma_i$ = variable amplitude stress range;
N_T = total number of cycles (= $\sum_i n_i$);
m = slope of corresponding constant amplitude S-N line.

The effective stress range of the variable stress spectrum, $\Delta\sigma_{ef}$ from the above equation is equal to the root mean square (RMS) if m is taken as 2. If m is taken as the slope of the constant amplitude S-N curve for the particular detail under consideration, the equation is equivalent to Miner's law, and for most structural details, m is about 3. The effect of the mean stress in each cycle of the variable-amplitude stress spectrum on the effective stress range is not considered in the above equation. If the fatigue damage model based on continuous damage mechanics (CDM) (Li et al., 2002) is applied instead of the Miner's fatigue model, the effective stress range so that is obtained as follows:

$$\Delta\sigma_{ef} = \left\{ \sum_i \frac{n_i}{N_T} \left[(\Delta\sigma_i + 2\sigma_{mi})\Delta\sigma_i \right]^{\frac{\beta+3}{2}} \right\}^{\frac{1}{\beta+3}} \tag{7}$$

The identification of coefficients in the model needs the measurement of damage derived by means of fatigue test when strain amplitude is controlled and the Woehler curve in the case of uniaxial periodic fatigue test. The model identification and the comparison of the theoretical results with the experimental data have been carried out for the effective stress range by Miner's and CDM fatigue damage model methods. The experimental data of fatigue carried out by Nielsen et al. (1997) on plate specimens with transverse attachments is applied. Since traffic loading is the dominant variable amplitude loading on steel bridge decks, the variable amplitude loading used in their fatigue tests had been determined from strain gauge measurements on the orthotropic steel deck structure of an existing bridge in Denmark.

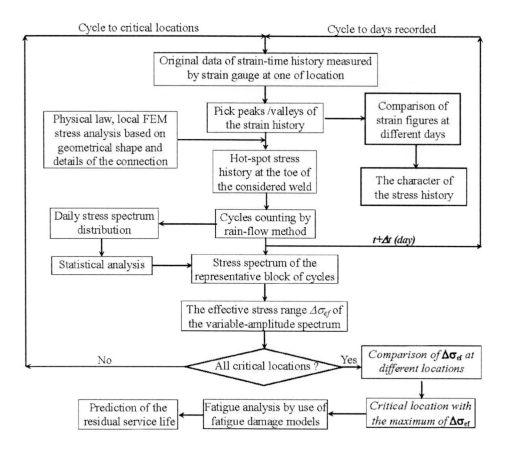

Figure 4. A flow chart of the approach for fatigue assessment

The effective stress range of the variable-amplitude spectrum at a given location is a quantitative description of fatigue stress and fatigue resistance (class of the welded details, fatigue stress limit, etc.) for the member under consideration. Therefore, a value or a normalized value of the effective stress range can be considered as a representative value of fatigue behavior at the location to be considered. From this point of view, the fatigue analysis of bridge-deck sections should be carried out only for the critical location on the section where the effective stress range reaches comparable large value at the section. As shown in Figure 4, fatigue analysis is carried out for the critical location instead of being done for all locations with a strain gauge. The most critical location is detected by comparing the value of the effective stress range and details near the location of the relevant strain gauge.

Application to the Tsing Ma Bridge

In this section, the approach proposed above is applied to the fatigue analysis of a bridge deck-section of Tsing Ma Bridge. Figure 5 shows locations of all strain gauges, including two sets of rosette strain gauges and 42 linear strain gauges, set in the cross frame at CH 24662.5 (HK HYD 1998). The location of each strain gauge in the cross frame can be better found from Table 8 together with Figure 5.

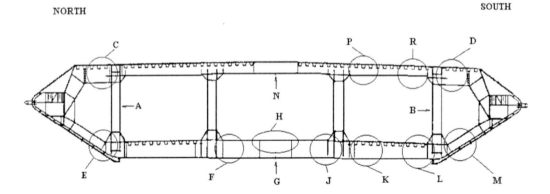

Figure 5. Strain gauge locations in the cross frame at CH 24662.5

The distribution of the effective stress range at each location of strain gauges is listed in Table 8. The effective stress range at each location of linear strain gauge is calculated by using Eq. 7, in which the coefficient $(\beta+3)$ is simply taken as the same value for all details, and the calculation for the two rosette strain gauges are not included. The distribution of the effective stress range is given by a normalized ratio of $\Delta\sigma_{ef}/\Delta\sigma_0$, where $\Delta\sigma_0$ is the value of effective stress range at the location of the strain gauge "SSTLN-01." It is observed that several critical locations of fatigue are found at "SSTLN-01" on the Kowloon bound and "SSTLS-05," "SSTLS-12," "SPTLS-09" and "SPTLS-14" on the Airport bound, respectively, where the effective stress range takes its greater value than other locations.

It can be seen that fatigue damage assessment for the deck section CH 24662.5 of the Tsing Ma Bridge should be carried out for the weld details near to the strain gauges "SSTLN-01" on the Kowloon bound and "SSTLS-05," "SSTLS-12," "SPTLS-09" and "SPTLS-14" on the Airport bound, respectively. These strain gauges are set respectively on the top chord of the north main longitudinal truss, south track 2, rail-way beam single 2, the south lower bracing of the cross frame and rail-way beam pair 1. The effective stress range reaches its maximum at "SPTLS-09," which is located at the south lower bracing of the cross frame under the railway. This result suggests the accumulative fatigue of the deck section to be dominated by the rail traffic loading.

Table 8. Distribution of effective stress range and location of strain gauges on the bridge-deck at CH 24662.5

Gauge No	Location	$\Delta\sigma_{ef}/\Delta\sigma_0{}^*$	Gauge No	Location n	$\Delta\sigma_{ef}/\Delta\sigma_0{}^*$	Gauge No	Location n	$\Delta\sigma_{ef}/\Delta\sigma_0{}^*$
SSTLN-01	A	1.00	SPTLS-04	N	---	SSTLS-01	B	0.33
-02	E	0.54	-05	N	---	-02	P	0.58
-03	A	0.54	-06	P	0.15	-03	R	0.35
-04	F	0.48	-07	R	0.13	-04	J	0.38
-05	F	0.51	-08	G	---	-05	J	1.52
-06	F	0.34	-09	G	1.73	-06	J	0.14
SPTLN-01	A	0.58	-10	K	0.36	-07	K	0.47
-02	A	0.67	-11	L	0.39	-08	L	0.39
-03	A	0.54	-12	B	0.70	-09	B	---
-04	E	0.27	-13	M	0.36	-10	M	0.11
-05	A	0.66	-14	H	1.12	-11	H	0.30
SPTLS-01	B	0.39	-15	H	0.77	-12	H	1.11
-02	B	0.99	-16	H	0.66	-13	H	0.65
-03	B	0.78	-17	H	0.89	-14	H	0.99

* $\Delta\sigma_0$ is the effective stress range at the location of the strain gauge "SSTLN-01"
--- Strain data failed to be recorded here

The critical location of the deck section will be one of those locations with comparable big value of the effective stress range, in the section under consideration, which will be one of locations of "SSTLN-01," "SSTLS-05," "SSTLS-12," "SPTLS-09" and "SPTLS-14". The most critical location is detected by comparing the value of the effective stress range at these locations and details near the relevant strain gauge. The assessment of the fatigue damage and prediction of residual service life can be made by using Miner's law or fatigue damage model based on CDM, which should be carried out for these critical locations. As an example of assessing fatigue damage and prediction of service life by Miner's law or nonlinear fatigue damage model based on CDM, fatigue damage at the weld near to strain gauge "SSTLN-01" is calculated and shown in Table 9. Values of predicted total service life are also given.

Table 9. Calculated Results by CDM model and Miner's rule

	Fatigue damage		Service life (Years)	
Strain gauge location	CDM	Miner's	CDM	Miner's
SS-TLN-01	0.0044	0.0139	159.3	121.3

It can be seen from the Table 9 that the fatigue damage calculated by Miner's rule is greater than that by CDM model, and so lesser values of predicted service life are obtained by Miner's rule.

USE OF MULTI-SCALE MODELING IN FATIGUE ASSESSMENT

Commonly used fatigue design rules are based mainly on the data generated from tests on small-scale specimens incorporating the welding detail of interest subjected to unidirectional loading. Fatigue test results can, therefore, be expressed in terms of the nominal applied stress in the region of the test detail, and the same stress is usually specified for use with the design S-N curves. An alternative approach, which is widely applied and is already established for tubular connections, is based on the hot-spot stress, where the stresses are near the fatigue failure of the weld toe. It can be seen from Table 1 that the hot-spot stress method will be adopted in the fatigue assessment. The hot-spot stress method relates the fatigue life of complex steel structure joints with the hot-spot stress, rather than the nominal stress. It considers the uneven stress distribution in the complex steel structure joints and incorporates geometry influences and excludes the effects related to the weld configuration and weld toe. The hot-spot stress is the maximum stress occurring in the complex steel structures, where cracks are most likely initiated.

In practical numerical analysis of such a large structure at the global level, it often treats one component as one element with the lower dimensional type, and the higher dimensional type elements for local details are not represented in such a global model. Hence, the hot-spot stress analysis procedure is divided into two steps: the global structural analysis is first carried out using a global model to determine various critical locations; and the detailed local models of desired critical regions are then constructed to obtain the corresponding hot-spot stress situations and, subsequently, to obtain the Stress Concentration Factor (SCF), which is defined as the ratio of the hot-spot stress value with that of the nominal stress. With the SCF value obtained, the hot-spot stress can be obtained by multiplication of the nominal stress with the SCF value. The SCF value of a welded joint is commonly obtained by experimental numerical methods. The welded joint of the Tsing Ma Bridge is very big and complex; it is almost impossible to obtain the SCF value in a laboratory. Hence, the numerical finite element method is chosen as an alternative to obtain the SCF value, which is widely used in the determination of the SCF value in tubular welded joints (CIDECT, 2001). In order to apply the hot-spot stress approach to the fatigue evaluation of the Tsing Ma Bridge, it is necessary to carry out the analysis in two steps: the global structural analysis is first carried out using a global model to determine various critical locations; and the detailed local models of desired critical regions are then constructed to obtain the corresponding hot-spot stress situations and, subsequently, to obtain the Stress Concentration Factor (SCF), which is the basis of the fatigue status assessment. However, because of the recent development of multi-scale modelling techniques, it will be much more desirable to combine the reduced or lower dimensional element types with higher ones in a single model to simultaneously consider the structural responses at different spatial scale levels, taking into account the interactive effects.

Strategies of Multi-Scale Modeling

In general, structural analysis is usually carried out at different scale levels for different purposes. As listed in Table 10 (Li et al., 2007), for a large structure, the size of the model beam element is usually at the level of meter for global structural analysis. Meanwhile, the beam element is also adopted at the centimeter level of component scale for nominal stress analysis. Furthermore, the scale of material points, usually modeled with shell or solid elements, is commonly less than the level of a millimeter for the local detailed hot-spot stress analysis. The major difficulties existing for each modeling scale method are the strict limitations on structural analysis on their own scale level instead of considering the other scales. Although some techniques such as transition element and mesh refinement have been applied in the modeling process and researchers have tried to make connections between different scale levels, they are actually still at the same scale level without real solutions to the problem.

Table 10 Physical scales and available theories for numerical analysis for long-span bridges

Scale Level	Representative Volume Element (RVE) for analyses	Objectives of Analyses	Available Theory
Global Scale (Up to km)	5~10 m	Internal Forces	Theory of Structures
Component Scales (Meters)	5~10 cm	Nominal Stress	Mechanics of Materials and Theory of Elastic-Plasticity
Material Scales (Microns)	1~5 mm or less	Hot-spot Stress or Effective Stress	Meso-mechanics, Damage Mechanics

The objective of multi-scale modeling is to construct the global model at the structural level and incorporate the local details of constitutional relationships for analytical purposes, and the local details could even be modeled layer by layer if necessary as shown in Figure 6. All the modeling process should be concurrent, and characteristics of corresponding scale level must be taken into account at the same time. In particular, with no loss of generality, the linear analysis could be performed at the global model of structural elements, and the corresponding linear responses can be obtained while nonlinear analysis must be considered to simulate the nonlinear behavior in local detailed models of higher levels.

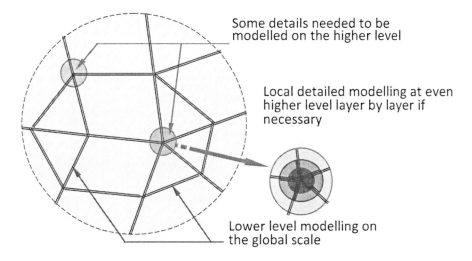

Some details needed to be modelled on the higher level

Local detailed modelling at even higher level layer by layer if necessary

Lower level modelling on the global scale

Figure 6. A sketch of description on multi-scale modeling

Mixed Dimensional Coupling Methods

Conventionally, the element types of different dimensions are coupled together using rigid links, known as "kinematic coupling." This is usually considered as the most basic and widely used method of coupling (Adams and Askenazi, 1999). As the constraints imposed by kinematic coupling are usually calculated as a function of the nodal coordinates, considerable stress disturbances are induced at the transition. The relatively large induced errors impose strict limitations on the application of kinematic coupling on practical engineering problems. Apart from several kinds of specifically developed transition elements, there are other coupling ways provided in commercial software package. There has been documented and implemented a general transition element in ABAQUS (HKS, 1998), which couples a reference node to a group of coupling nodes and distributes forces and moments at the reference node with a weight factor assigned. However, the correct values of these weights usually are very difficult to determine and currently are determined manually by the user with so many uncertainties and factors in the weight values estimation.

A multi-point constraint equation is an equation that defines a relationship between sets of displacements within a finite element model. Mathematically, there are two main methods of implementing these constraints, using Lagrange multipliers or using a penalty function method. The Lagrange multiplier method enforces constraints exactly and the penalty function method imposes constraints approximately. The scheme was proposed by McCune (1998) and further developed by Monaghan (2000), who utilized an outcome of Reissner's bending theory of elastic plates and showed that proper connections between plate and beam elements, and plate and solid elements, could be achieved via multipoint constraint equations evaluated by equating the work done on either side of the dimensional interface. The solution could be arrived at by introducing the assumed variation of the stresses, given by the appropriate beam, plate or shell theory, over the cross-section of the interface. This procedure is general, and as long as the stress distribution due to any given loading can be determined at each interface, beam-solid (Monaghan et al., 1998), beam-shell (Monaghan, 2000) or shell-solid (Monaghan, 2000 and Da´vila, 1994) coupling can be achieved for transitions with arbitrary shapes of cross-sections. Mixed dimensional coupling using constraint equations has been shown to give good results (Monaghan et al., 1998).

Compared within the mixed dimensional coupling methods mentioned above, the scheme of constraint equations is eventually adopted to perform the connections between elements of different dimensions. The situation of coupling beam with shell elements would be categorized as four simple loading scenarios, namely axial forces, bending moment, torsion and shear forces, since the different types of constraint equations would be developed for the structure under the different internal forces. The work generated by each type of element is firstly equated, and the expression of equating work would then be transformed to nodal displacements of different types of elements by using the shape function. Once the connection interface pattern between two element types is determined, the coefficients of constraint equation would be easily calculated by a specifically developed FORTRAN program. The corresponding constraint equations are conveniently derived and implemented by the command EQUATION in ABAQUS. The application of mixed dimensional modeling can be more beneficial when it is used in conjunction with substructuring, as shown in Figure 7 below.

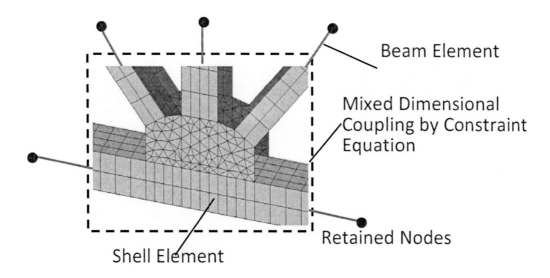

Figure 7. FE joint model by mixed dimensional coupling

Multi-Scale Model of TMB

A finite element global structural model embodying the properties of almost all the structural elements with more than 7300 nodes and 19000 elements was developed, including longitudinal trusses, cross-frames, deck plates, tower beams, main cables, hangers, etc. The FE model of the a deck section of the TMB is shown in Figure 8. The bridge deck (Figure 8b) is a hybrid steel structure consisting of vierendeel cross-frames supported on two longitudinal trusses acting compositely with stiffened steel plats that carry the upper and lower carriageways. In the developed FE model (Figure 8d) based on design drawing, longitudinal trusses in the bridge deck, comprising chords, posts and diagonal bracings, were modeled as 3-D, two-node iso-parameteric beam elements having six degree of freedom at each node. Bending, torsional and axial force effects were all considered in each space beam element. The section details were also embodied in the corresponding elements: the top and bottom chords were modeled as space beam elements with box sections; vertical posts and diagonal bracings were represented by I-beam elements. The main cables and hangers, generally simulated as truss elements, were modeled as two node space beam elements with circular cross-sections. Highway pavements were modeled by 3-D, four-node, doubly curved shell elements with reduced integration, while railway beams were made up of two inverted T-beams welded to flange plates, and each of them was represented by continuous beam elements with I-section. The Tsing Yi and Ma Wan towers, which are made of reinforced concrete supported by hard foundation, were modeled as rigid bodies. All of the corresponding element properties were listed in details in Xu et al. (1997).

On the basis of the above developed TMB global model, the concerned details of joints should be considered and merged into the global model to construct the multi-scale model. In the global model (Figure 9a), the component was modeled as space beam element, while the shell element would be required to simulate the joint details for the deteriorating process. There were six standard elements identified and modeled in the typical longitudinal truss section of TMB, and the joints were repeated along the longitudinal direction of the bridge. Based on the mixed dimensional method by constraint equations and substructuring technique, any concerned detailed joint (Figure 9c), which had been identified as the vulnerable regions to fatigue damage, could be modeled as substructure and merged into the global model to obtain the multi-scale model (Figure 9b). In the concerned local detailed model, the components such as the upper and lower orthotropic steel decks, the pavement structures and the aerodynamic stainless steel sheets had been considered as equivalent simpler models in the global model. It is understood that if these components were not considered properly, the bridge so modeled would be much more flexible than real structure, and the computed results due to train-induced displacements would be larger than that of the measured results. Therefore, these components have to be modeled on the basis of real structure details, if necessary. The multi-scale FE model of TMB was eventually developed, and it could be updated using the online monitoring data and field tests. Furthermore, the local damage evolution and its influence on the structural response can be considered and simulated in the modeling. The developed model, as a discrete physical and mathematical representation of the bridge, will serve as a baseline FE model for multi-scale analysis to provide the global structural stress distribution and local damage evaluation accounting for deterioration at critical locations.

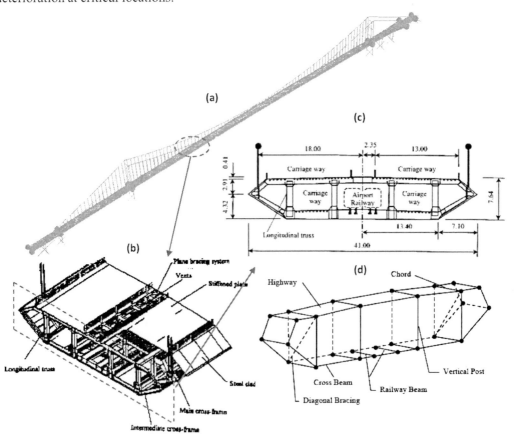

Figure 8. The FE model of TMB deck section

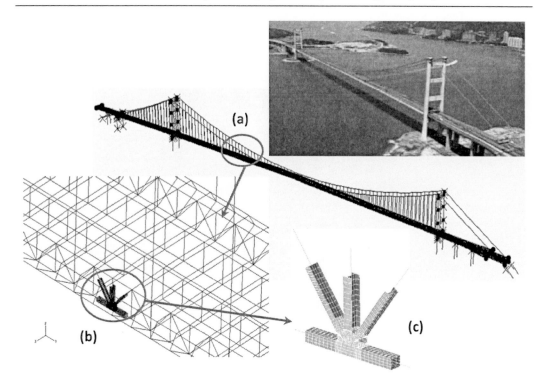

Figure 9 Multi-scale model of TMB: (a) Conventional model with beam and shell elements; (b) Multi-scale model; (c) Local detailed model

Verification of the TMB Multi-Scale Model

Before applying the multi-scale model to hot-spot stress analysis for fatigue assessment, the modal properties of developed model under free vibration and static response would need to be verified. Meanwhile, the comparison of the computed results with those of original global model and online output data would also be carried out to verify the efficiency and accuracy of the developed multi-scale model.

Dynamic Characteristics

The first few free vibration frequencies were measured by the Hong Kong Polytechnic University (HK PolyuU) and the Hong Kong Highways Department (HK HyD) by field test (Wong, 2003) as listed in Table 11, which presents the computed values of original structural model and developed multi-scale model for comparison. Corresponding modal analysis criterion values (MAC) Modal Mass Participation Factors (MMPF) are also shown in the table. The calculated values are used to compare with the mean experimental values to obtain the maximum relative difference is 3.25%, which indicates that the frequencies calculated by the multi-scale model agree well with the measured and the analyzed ones.

Table 11. Few Free Vibration Frequencies Comparison of TMB Original Model and Multi-scale Model

Mode	Experimental Results			Calculated Results by FEM					
	HK PolyU	HK HyD	Mean	Original Global Model		Multi-scale Model		MAC	MMMF
				Value	% Diff.	Value	% Diff.		
Lateral									*
First	0.069	0.068	0.0685	0.0683	-0.292	0.0683	-0.292	9.92	1.3324
Second	0.164	0.158	0.161	0.160	-0.621	0.160	-0.621	9.90	-0.0296
Vertical									**
First	0.113	0.113	0.113	0.116	+2.65	0.116	+2.65	9.86	-0.035
Second	0.139	0.138	0.1385	0.143	+3.25	0.143	+3.25	9.77	0.989
Torsion									***
First	0.267	0.244	0.2555	0.255	-0.196	0.255	-0.196	9.78	52.2
Second	0.320	0.266	0.293	0.299	-2.05	0.299	-2.05	9.80	32.5

* : Z-Component,**: Y-Component, ***: X-Rotation

Static Response by Vertical Displacement Influence Line

Apart from the dynamic characteristics with respect to natural frequencies, the developed multi-scale model should be further verified with respect to static and dynamic responses. The data of vertical displacement influence line at the locations of A, B and C (Figure 10), where GPS level stations are installed in the bridge deck are collected from the field test performed under a typical train loading. As for the numerical analysis, the simulated train loading had been properly calculated by Chan et al. (2007) and then applied to the multi-scale model to obtain the responses at those sensor-installed locations. There are relatively large differences between the measurement data and predicted ones as shown on right side of Figure 10.

The large discrepancies could be mainly due to three reasons:

i. Amplitude of train loading—The simulated train loading is estimated on the basis of the situation of full loading that is larger than the real experimental value, causing the larger vertical displacement

ii. The possible time-shift—During the simulation analysis, time zero is assumed when the train begins to pass through the bridge. However, the selected measurement data is just a portion of the whole time-history record without clearly knowing when the starting time is.

iii. The X-axis scale—The train is assumed to be traveling across the bridge at a constant speed of 30 m/s (≈108 km/hr), while in practice it may not be so. The differences of train speed between the simulation and real train result in the different scales along the X-axis.

In order to compare the calculated values at these locations with the measurement data with respect to the vertical influence line, three factors considering each of the above-mentioned three reasons were calculated in an attempt to tune the factored values to fit the configuration of measurement data that corresponds to GPS location A. Once the three factors were determined, the coefficients were multiplied to the other two cases, and the original calculated data was then calibrated as shown in Figures 5(b) & (c). It can be easily seen that all the calibrated data are now in better agreement with the measurement than the original prediction values, indicating that the developed multi-scale model is reliable and accurate in static response analysis of bridge.

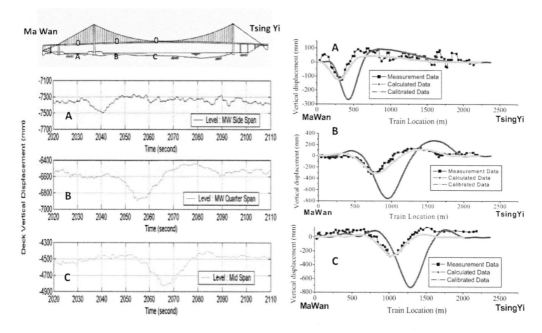

Figure 10. Comparison of calibrated values with calculation and measurement at GPS Locations A, B & C

Stress Concentration Factor

The TMB carries both highway loading and train loading. Theoretically, its dynamic response is caused by the combination of the both loadings. However, the highway loading is very complicated, and there are still many uncertainties in formulating the equivalent truck-loading model. The train has the fixed axle spacings and known axle loadings, and particularly it often runs on a fixed schedule. The train-loading model is, therefore, more reliable than the truck-loading model. Hence, for simplicity, only the dynamic response under the train loading was taken into account in this paper. The train-loading and corresponding calculation procedures and the detailed locations of strain gauges in bridge deck had been preliminarily studied. A comparison between the computed result of multi-scale model and the measured data of strain-time history of strain gauges was carried out (Chan et al., 2005). The strain record data should be transferred to stress and also calibrated for comparison with the computed values; i.e., the selected stress block should be moved to locate at the same time as the computed data, and the measured stress value ought to be equal to the computed ones at zero time point.

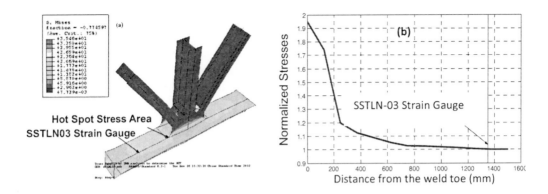

Figure 11. Stress distribution of typical joint at longitudinal truss: (a) Stress contour at the joint details; (b) Normalized stresses along the truss

On the basis of the plot of strain history data, there is only a stress cycle selected from the data within the passage of a train. However, for the measured data, the small stress cycles could be neglected as only the main stress cycle will be considered for the high-cycle fatigue damage analysis. The computed response stress history curve agrees with the measured data, to some extent. Consequently, the dynamic responses of the detailed joint as a substructure could be simultaneously obtained using the multi-scale model. The stress distribution contour and normalized stress along the longitudinal truss of the typical detailed intersection joint is shown in Figure 11, when it reaches its maximum values of stress history response under the train loading. The hot-spot just locates at the vicinity of weld toe, and the normalized stress curve is shown in Figure 11-b. It shows that the stress increases from the location of the stain gauge to the hot-spot stress area. Also the corresponding hot-spot stress value and SCF value could be accordingly obtained and used for carrying out the subsequent fatigue evaluation procedure.

CONCLUSION

In recent years, various sophisticated Structural Health Monitoring (SHM) systems—albeit with some limitations—have been installed overseas for significant bridges to ensure their normal operation and safety. A structural health monitoring system, named as "Wind And Structural Health Monitoring System," or WASHMS, has been installed on the six cable-supported bridges in Hong Kong. Structural Health Monitoring should be composed of two main components: Structural Performance Monitoring (SPM) and Structural Safety Evaluation (SSE). As an example to illustrate how the WASHMS could be used for structural performance monitoring, the layout of the sensory system installed on the Tsing Ma Bridge is briefly described. The system could be used to monitor the structural health status of the bridge under normal conditions and to evaluate structural degradation as it occurs. To demonstrate the two broad approaches of structural safety evaluation—Structural Health Assessment and Damage Detection—three examples in the application of SHM information are presented. These three examples can be considered as pioneer works for the research and development of the structural diagnosis and prognosis tools required by the structural health monitoring for monitoring and evaluation applications

Regarding the damage detection approach, this article presents the numerical simulation results on using the auto-associative neural networks for the detection of damage occurrence in the three cable-supported bridges in Hong Kong, namely the Tsing Ma Bridge, Kap Shui Mun Bridge and Ting Kau Bridge. Since no structural model is required, the novelty detection method in terms of the auto-associative neural network provides an innovative technique for structural damage alarming. This method should be applicable to structures of arbitrary complexity. It is concluded from the simulation studies that when a "normal" random fluctuation with 0.005 variance (maximum error 1.5%) is considered, the novelty indexes are able to unambiguously alarm the damage states if the maximum frequency change ratio is greater than 1.0% for the Tsing Ma Bridge and Ting Kau Bridge and 1.3% for the Kap Shui Mun Bridge. The occurrence of damage cannot be flagged if the maximum frequency change ratio is less than 0.3% for the Tsing Ma Bridge, 0.25% for the Kap Shui Mun Bridge and 0.4% for the Ting Kau Bridge. When the maximum frequency change ratio is between the two critical percentages of the corresponding bridge, e.g., 0.3% and 1.0% for the Tsing Ma Bridge, the damage is just detectable with a weak alarming signature.

Regarding the structural health assessment approach, a proposed strategy for evaluating fatigue damage and detecting possible critical locations of local fatigue on bridge-deck sections are presented. This strategy allows fatigue assessment to be carried out by taking full advantage of the online monitoring data. It can be seen that strain histories recorded by structural health monitoring system for the bridge under actual traffic loading can be modeled by a repeated block cycles in which the standard block, defined as a representative block of daily strain cycles, is repeated everyday. The effective stress range for the variable-amplitude stress spectrum due to traffic load can then be used to evaluate the fatigue strength of the bridge-deck section at different locations of strain gauges, by which the possible location of critical fatigue failure can be primarily determined. Based on the analysis in this work, the critical location of the deck section CH 24662.5 of the TMB will be one of those locations with greater value of the effective stress range. The locations of "SSTLN-01," "SSTLS-05," "SSTLS-12," "SPTLS-09" and "SPTLS-14" should be paid special attention in nondestructive inspections and fatigue damage evaluations.

An important concept on the adoption of the multi-scale modeling strategy for the structural health monitoring of the long-span bridges are also introduced as the third example. As an important case study, the multi-scale model of TMB is developed on the basis of the FE original model. The multi-scale modeling is based on the mixed dimensional coupling method by constraint equations generated at the interface between different types of finite elements on the basis of equating work. The constraint equations could be derived using developed computer programs and easily implemented in a general purposed software like ABAQUS. Combined with the substructuring technique, the constraint equations of detailed joints connections are derived and implemented to the longitudinal truss section of TMB to obtain the corresponding multi-scale model on the basis of original FE model. The comparison of the first few natural frequencies, vertical displacement influence line and dynamic response under the train loading by the developed multi-scale model with the original model and measured data, are performed and good agreements are obtained. As for the dynamic response under train loading, the hot-spot stress value and the corresponding SCF value of typical intersection joint were further obtained in one single step using the multi-scale model for the purpose of fatigue evaluation process demonstrating its efficiency when compared with the traditional two-step approach.

CONTINUING TO AUSTRALIA...

From the SHM experiences gained in Hong Kong, it is believed that further work of SHM research should be carried out in three directions, namely (i) System Development, (ii) Sensing Devices/Measurement Development and (iii) Applications, covering both the structural performance monitoring and structural safety evaluation components of SHM. The first author moved to Australia in 2007, and continues the research in SHM there. The research includes all these three directions. The work in the System Development direction aims to devise an effective system incorporating an optimized network of global and local sensors for a bridge structure and to investigate how the SHM data could be practically used to assess the health status of a bridge structure installed with a SHM system. The techniques in concurrent multi-scale modeling should be explored and further developed for SHM purpose. There are projects working on these. As it is necessary to develop and improve some innovative sensors or sensing technology for SHM, so in the Sensing Devices/Measurement Development direction, some new sensors and sensing technologies are under development, e.g., the development of acoustic emissions sensors for local damage detection and the development of new technology for measuring vertical displacements using fiber optics and CCD camera. Development of effective damage detection methods for real life applications is always an important objective of SHM researchers. Therefore, various methods for damage detection of civil structures are being explored under various projects. The following two chapters will describe some of the findings in these projects and other advances on structural health monitoring will be reported in later publications.

ACKNOWLEDGMENTS

The support of the Queensland University of Technology, Hong Kong Polytechnic University, the Research Grants Council of the Hong Kong SAR Government and the Highways Department of the Hong Kong SAR Government are gratefully acknowledged. Any opinions expressed or conclusions reached in the text are entirely those of the authors.

REFERENCES

Adams, V. and Askenazi, A. (1999) *Building better products with finite element analysis*: On Word Press, Santa Fe, New Mexico.

American Association of State Highway and Transportation Officials (1989) *Guide Specifications for Fatigue Design of Steel Bridges*, Washington, DC, USA.

American Association of State Highway and Transportation Officials (1990) *Guide Specifications for Fatigue Evaluation of Existing Steel Bridges*, Washington, DC, USA.

Catbas, F.N., Susoy, M. and Frangopol, D.M. (2008) "Structural health monitoring and reliability estimation: Long-span truss bridge application with environmental monitoring data." *Engineering Structures*, 30(9), 2347-2359.

Chan, T.H.T., Ko, J.M., Ni, Y.Q., Lau, C.K., and Wong, K.Y. (2000) "A Feasibility Study on Damage Detection of three Cable-Supported Bridges in Hong Kong using Vibration Measurement." *Proceedings of Workshop on Research and Monitoring of Long-Span Bridges*, April 2000, Hong Kong, pp 204-211.

Chan T.H.T, Zhou T.Q., Li Z.X. Guo L., (2005) "Hot-spot stress approach for Tsing Ma Bridge fatigue evaluation under traffic using finite element method." *J. of Structural Engineering and Mechanics*, 19(3): 261-279.

Chan, T.H.T, Yu, Y., Wong, K.Y., and Guo, L. (2007) "A Multi-scale Finite Element Model of Tsing Ma Bridge for Hot-spot Stress Analysis." *Proceedings of International Conference on Health Monitoring of Structure, Material and Environment (HMSME 2007)*, 16-18 October, Nanjing, PR China, 130-140.

Chang, S.P. and Kim, S. (1996) "Online structural monitoring of a cable-stayed bridge." *Smart Structures and Materials 1996: Smart Systems for Bridges, Structures, and Highways*, L.K. Matthews (ed.), SPIE Vol. 2719, pp. 150-158.

CIDECT (2001) *Design Guide for Circular and Rectangular Hollow Section Welded Joints Under Fatigue Loading*, TÜV-Verlag GmbH, Germany.

Da'vila C.G. (1994) "Solid-To-Shell Transition Elements for the Computation of Interlaminar Stresses." *Computing Systems in Engineering*, 5(2), 193–202.

Doebling, S.W., Farrar, C.R., Prime, M.B., and Shevitz, D.W. (1996) "Damage Identification and Health Monitoring of Structural and Mechanical Systems from Changes in Their Vibration Characteristics: A Literature Review." Los Alamos National Laboratory.

Farrar, C.R., and Worden, K. (2007) "An introduction to structural health monitoring." *Philosophical Transactions of the Royal Society A: Mathematical, Physical and Engineering Sciences*, 365(1851), 303-315.

Highways Department HK SAR Government (1998) Drawing No. P18/C9/E/TM/04/287, Contract No. HY/93/09, "Lantau Fixed Crossing, Electrical & mechanical Services." June 1998.

Hibbitt, Karlsson and Screnson, Inc. (1998) ABAQUS 5.8 Theory Manual.

Ko, J.M., and Ni, Y.Q. (1999) "Development of vibration-based damage detection methodology for civil engineering structures." *Proceedings of the First International Conference on Structural Engineering*, Kunming, China, pp. 37-56.

Ko, J.M., Sun, Z.G., and Ni, Y.Q. (2002) "Multi-stage identification scheme for detecting damage in cable-stayed Kap Shui Mun Bridge." *Engineering Structures*, Vol. 24, No. 7, 857-868.

Li, Z.X., Chan, T.H.T. and Ko, J.M. (2002) "Determination of Effective Stress range and its Application on Fatigue Stress Assessment of Existing Bridges." *International Journal of Solids and Structures*, Vol. 39, pp. 2401-2417.

Li, Z.X., Zhou, T.Q., Chan, T.H.T. and Yu, Y. (2007) "Multi-scale Numerical Analysis on Dynamic Response and Local Damage in Long-span Bridges." *Engineering Structures*, 29(7), 1507-1524

Mccune, R.W. (1998) *Mixed dimensional coupling and error estimation in finite element stress analysis.* PhD thesis, The Queen's University of Belfast.

Monaghan, D.J., Doherty, I.W., Mccourt, D., Armstrong C.G. (1998) "Coupling 1D beams to 3D bodies." 7th *International Meshing Roundtable* Sandia National Laboratories, Dearborn, Michigan, 285–293.

Monaghan, D.J. (2000) *Automatically coupling elements of dissimilar dimension in finite element analysis.* PhD Thesis, The Queen's University of Belfast.

Myrvoll, F., DiBiagio, E. and Hansvold, C. (1996) "Instrumentation for monitoring the Skarnsundet cable-stayed bridge." *Publikasjon–Norges Geotekniske Institutt*, No. 196, pp. 145-153.

Nellen, M.P., Anderegg, P., Bronnimann, R. and Sennhauser, U. (1997) "Application of fiber optical and resistance strain gauges long-term surveillance of civil engineering structures." *Smart Structures and Materials 1997: Smart Systems for Bridges, Structures, and Highways*, N. Stubbs (ed.), SPIE Vol. 3043, pp. 77-86.

Nielsen, J.A., Agerskov, H., and Vejrum, T. (1997) "Fatigue in steel highway bridges under random loading." *Rep. No. R 15*, Technical University of Denmark, Lyngby, Denmark.

Sohn, H., Farrar, C. R., Hemez, F. M., Shunk, D. D., Stinemates, D. W., Nadler, B. R., and Czarnecki, J. J. (2004) *A Review of Structural Health Monitoring Literature: 1996-2001.* Los Alamos National Laboratory.

The Hong Kong Polytechnic University (1998) *"State-of-the-Art in Vibration-Based Structural Damage Detection: Literature Review."* Report No. WASHMS-01, Department of Civil and Structural Engineering, 22 January 1998.

The Hong Kong Polytechnic University (1999) "Finite element modeling and modal sensitivity analysis of the Tsing Ma Suspension Bridge." *Report No. WASHMS-03(a)*, Department of Civil and Structural Engineering, 26 July 1999.

Worden, K. (1997) "Structural fault detection using a novelty measure." *Journal of Sound and Vibration*, 201, pp. 85-101.

Worden, K., Farrar, C. R., Manson, G., and Gyuhae, P. (2007) "The fundamental axioms of structural health monitoring." *Proceedings of the Royal Society of London, Series A (Mathematical, Physical and Engineering Sciences)* (pp. 1639-1664). R. Soc. London.

Wang, J. Y., Ko, J. M., and Ni, Y. Q. (2000) "Modal sensitivity analysis of Tsing Ma Bridge for structural damage detection." *Proceedings of the SPIE's Fifth International Symposium on Nondestructive Evaluation and Health Monitoring of Aging Infrastructure*, Newport Beach, California, USA, March 5-9, 2000, Vol. 3995, pp. 300-311.

Wong, K.Y. (2003) "Structural identification of Tsing Ma Bridge." *Transaction, The Hong Kong Institute of Engineers*, 10(1), 38-47.

Wong, K.Y., and Ni, Y.Q. (2009a) "Modular Architecture of Structural Health Monitoring System for Cable-Supported Bridges." Chapter 123 of the book: *Encyclopedia of Structural Health Monitoring*, edited by C. Boller, F.-K. Chang and Y. Fujino, John Wiley & Sons, Chichester, UK, Vol. 5, 2089-2105.

Wong, K.Y., and Ni, Y.Q. (2009b) "Structural Health Monitoring of Cable-Supported Bridges in Hong Kong." Chapter 12 of the book: *Structural Health Monitoring of Civil Infrastructure Systems*, edited by V. Karbhari and F. Ansari, Woodhead Publishing, Cambridge, UK, 371-411.

Xu, Y.L. Ko, J.M. and Zhang W.S. (1997) "Vibration Studies of Tsing Ma Suspension Bridge." *J. Bridge Engineering*, 2(4): 149-156.

In: Structural Health Monitoring in Australia   ISBN: 978-1-61728-860-9
Editors: Tommy H.T. Chan and D. P. Thambiratnam   ©2011 Nova Science Publishers, Inc.

Chapter 2

DAMAGE LOCALIZATION IN BEAMS AND PLATES USING VIBRATION CHARACTERISTICS

H.W. Shih*, David P. Thambiratnam≠ and Tommy H.T. Chan‡

ABSTRACT

Dynamic computer simulation techniques are used to develop and apply a multi-criteria procedure, incorporating changes in natural frequencies, modal flexibility and modal strain energy, for damage localization in beams and plates. Numerically simulated modal data obtained through finite element analyses are used to develop algorithms based on changes of modal flexibility and modal strain energy before and after damage and used as the indices for assessment of the state of structural health. The proposed procedure is illustrated through its application to flexural members like beams and plates under different damage scenarios, and the results confirm its feasibility for damage assessment.

Keywords: beam; plate; damage identification; damage localization, finite element method; modal analysis; modal strain energy; modal flexibility

INTRODUCTION

Bridges are usually designed to have long lifespans. During their service lives, changes in load characteristics, deterioration with age, environmental influences and random actions may cause local or global damage to these structures. Continuous health monitoring of structures will enable the early identification of distress and allow appropriate retrofitting to prevent potential catastrophic structural failures. In recent times, structural health monitoring (SHM) has attracted much attention in both research and development. SHM defined by Housner et al. (1997) refers to the use of in-situ, continuous or regular (routine) measurement and

* BEE, QUT, GPO Box, 2434 Q4001 (shihhoiwai@gmail.com)
≠ BEE, QUT, GPO Box, 2434 Q4001 (d.thambiratnam@qut.edu.au)
‡ BEE, QUT, GPO Box, 2434 Q4001 (tommy.chan@qut.edu.au)

analyses of key structural and environmental parameters under operating conditions, for the purpose of warning impending abnormal states or accidents at an early stage to avoid casualties as well as giving maintenance and rehabilitation advice. SHM encompasses both local and global methods of damage identification (Zapico and Gonzalez, 2006). In the local case, the assessment of the state of a structure is done either by direct visual inspection or using non-destructive evaluation (NDE) techniques such as those using acoustic emission, ultrasonic, magnetic particle inspection, radiography and eddy current. A characteristic of all these techniques is that their application requires a prior localization of the damaged zones. The limitations of the local methods can be overcome by using vibration-based (VB) methods, which offer a global damage assessment. Health-monitoring techniques based on processing vibration measurements basically relate two types of characteristics: the structural parameters (mass, stiffness, damping) and the modal parameters (modal frequencies, associated damping values and mode shapes). As the dynamic characteristics of a structure, namely natural frequencies and mode shapes, are known to be functions of its stiffness and mass distribution, variations in modal frequencies and mode shapes can be an effective indication of structural deterioration. Deterioration of a structure results in a reduction of its stiffness, which causes the change in its dynamic characteristics. Thus, damage state of a structure can be inferred from the changes in its vibration characteristics (Doebling et al., 1996). Usually, there are four different levels of damage assessment (Rytter, 1993): damage detection (Level 1), damage localization (Level 2), damage quantification (Level 3), and predication of the acceptable load level and of the remaining service life of the damaged structure (Level 4). There has been a relatively large amount of literature dealing with single-damage scenarios, especially in simple structural elements, but limited number for multiple-damage scenarios. The methods based only on changes in frequencies and mode shapes are limited in scope and may not be useful in some real life application situations. It is observed that changes in natural frequencies alone may not provide enough information for integrity monitoring. It is common to have more than one damage case giving a similar frequency-change characteristic ensemble. In the case of symmetric structures, the changes in natural frequency due to damage at two different symmetric locations are exactly the same. Alternatively, no changes in the mode shapes could be detected if the mode had a node point at the location of damage (Farrar and Cone, 1995). There is thus a need for a more comprehensive method of damage assessment in structures.

Fast computers and sophisticated finite element programs have enabled the possibility of analyzing hitherto intractable problems in structural engineering while simplifying the analyses of other problems. In this chapter, dynamic computer simulation techniques are used to develop and apply two non-destructive damage detection parameters, along with changes in natural frequencies, for damage identification in flexural members. These parameters are the modal flexibility and the modal strain energy, which are based on the vibration characteristics of natural frequencies and mode shapes and their variations with the health states of the structure. The application of the approach is demonstrated through numerical simulation studies of beams and plates with different damage scenarios.

VIBRATION-BASED DAMAGE IDENTIFICATION

There are a large number of publications on the damage assessment of structures using numerical, experimental and theoretical procedures. Most of the research pertains to damage assessment in structural members such as beams and plates, while some treat a bridge or a building frame. Some of the most recent and relevant research is reviewed here, especially those on vibration-based damage assessment methods.

An extensive review of methods that directly use the vibration characteristics such as frequencies, mode shapes, damping, etc., to detect structural damage has been carried out by Araújo dos Santos et al. (2008). They also review model updating methods and infer that the vast amount of exiting knowledge can be applied to all kinds of structures and damages types. Fang and Perera (2009) used two damage indices—power mode shape curvature and power flexibility, derived from power mode shapes obtained from signal power spectral densities to successfully detect structural damage in structures.

Damage assessment in structural elements and frames has received considerable attention. The use of vibration-based parameters in assessing the effect of repairs to reinforced concrete beams was studied by Baghiee et al. (2009). They carried out vibration tests on healthy, damaged and repaired reinforced concrete beams and assessed the capability of different parameters such as frequency, modal assurance criterion, coordinate modal assurance criterion and modal curvature in assessing the repair. El-Ouafi et al. (2009) proposed a damage diagnosis method for beams and slabs, which involved the calculation of an "error" based on vibration data relative to the current and reference states of the structure and demonstrated its ability to assess moderate degrees of damage. Ge and Lui (2005) proposed a method for determining the location and severity of damage in three structures, a beam, a frame and a plate, using the corresponding undamaged structure's stiffness and mass properties and the damaged structure's vibration properties. They pointed out that good results can be expected if suitable finite element models and reliable vibration data are available. A virtual-energy-based approach that estimates reduction of the strain energy of vibration modes to assess the damage of a reinforced concrete beam was proposed by Petryna et al. (2002). Finite element and Monte-Carlo techniques are used to solve the Eigenvalue problem with respect to randomly generated stiffness and mass matrices and to determine the damage. The success of the method is demonstrated by comparing numerical and experimental results. Damage assessment in a reinforced concrete frame was treated by Curadelli et al. (2008) by evaluating the damage-sensitive damping coefficient using wavelet transforms. They carried out vibration measurements on the structure under ambient or controlled excitation and correlated damping with damage.

Damage in bridges has also received some attention using vibration-based methods and other techniques. Ren and Sun (2008) adopted the wavelet transform combined with Shannon entropy to detect structural damage of a bridge model from measured vibration signals. Some damage features such as wavelet entropy, relative wavelet entropy and wavelet-time entropy are defined and investigated to detect and locate damage. The procedure is illustrated by a numerically simulated case and two laboratory test cases and the merits of these damage features are discussed. The integrity of shear connectors in a 1:3 scaled bridge model was assessed by Xia et al. (2007) adopting various vibration-based damage identification methods for different damage scenarios in the connectors. The results showed that a local approach was able to detect all the damage successfully and consistently and as this approach did not need any reference data for the structure, it could be applied to the prototype bridges. Modal parameters from ambient vibration, together with neural networks, were used by Lee and Yun (2006) to determine damage location and severity of steel girder bridges. The effectiveness of the proposed method was demonstrated through numerical analysis and experimental testing of a simply supported bridge model with multiple girders. Petryna et al. (2006) carried out a geometrically and physically nonlinear finite element analyses to simulate and assess structural damage and lifetime of an arch bridge. The degradation of structural performance was simulated and design, material and damage parameters were considered as random values to account for possible uncertainties. The numerical results on structural and material state of the bridge after 54 years of service life were predicted and compared with the actual state observed before demolition through vibration tests and examination of different structural elements, respectively.

Optimization techniques in combination with modal data have been used in damage assessment. Perera et al. (2007) presented the framework for damage identification problems using measured modal data formulated as a multi-objective optimization problem and describe the methodologies for solving such problems and compare their merits. Guan and Karbhari (2008) introduced an improved damage detection method in which the damage index does not rely on numerical differentiation, but is calculated using only modal displacement and modal rotation. A penalty-based minimization approach is used to find the unknown modal rotation using sparse and noisy modal displacement measurement. The feasibility of the procedure is confirmed through numerical simulation and experimental validation.

Forced vibration and wave propagation techniques have also been used for damage assessment. Zacharias et al. (2008) proposed a method for fatigue crack detection in beam-like structures based on nonlinear vibration. The nonlinear dynamic behaviour of a cantilever beam with a fatigue crack under harmonic excitation is investigated both theoretically and experimentally, and a relationship for crack size was established for damage assessment. Fernandes et al. (2008) compared the influence of damage on the vibration wave propagation features of a slender Euler-Bernoulli beam. In the vibration framework, the damage is assessed by considering changes in the reduced flexibility matrix of the structure. In the wave propagation framework, the presence of damage is perceived by an early echo output. A slender aluminium beam with an imposed damage scenario is considered for the numerical analysis. Both approaches showed to be sensitive to damage and hence can be used for identification. Peng et al. (2008) developed a simple model to track non-linear phenomenon in cracked structures under sinusoidal excitation. This is then applied to analyse the crack-induced non-linear response of a finite element model of a beam to successfully indicate the existences and the sizes of cracks.

As shown from the above review, a number of methodologies have been used to identify, locate and estimate the severity of damage in structures using numerical simulations (Lee et al., 2004). The most common vibration-based (VB) damage detection techniques include those which utilize changes to mode shapes, modal curvatures, flexibility curvatures, strain energy curvatures, modal strain energy, flexibility and stiffness matrices. Other techniques, such as numerical model updating and artificial neural network, have also been incorporated into VB damage detection methods. It is noticed that those methods utilizing mode shapes are the better developed in terms of the ability to identify, locate and estimate the severity of damage, especially in the field applications. The advantage of using the modal parameter such as modal flexibility is that the flexibility matrix is most sensitive to changes in the lower-frequency modes of the structures due to the inverse relationship to the square of the natural frequencies (Huth et al., 2005). Therefore, a good estimate of the modal flexibility can be made with the inclusion of the first few frequencies and their associated mode shapes. The advantage of using modal strain energy parameter is that only measured mode shapes are required in the damage identification without knowledge of the complete stiffness and mass matrices of the structure (Stubbs et al., 1995; Cornwell et al., 1999). With the mode shapes of the first few modes and their corresponding derivatives, this proposed algorithm can accurately predict multiple damage locations (Choi et al., 2008). By using both modal strain energy and modal flexibility simultaneously, the shortcomings from each individual parameter can be compensated as the two parameters complement each other and improve accuracy and reliability of detection results for single and multiple damages. The objective of this chapter, therefore, is to evaluate the feasibility and capability of simultaneously using these two proposed parameters for multiple damage localization in flexural members such as beams and plates.

THEORY

Modal Flexibility Matrix

The modal flexibility matrix includes the influence of both the mode shapes and the natural frequencies. It is defined as the accumulation of the contributions from all available mode shapes and corresponding natural frequencies. The modal flexibility matrix associated with the referenced degrees of freedom can be established from Eq. (1) found in Huth et al. (2005).

$$[F] = [\phi][1/\omega^2][\phi]^T \tag{1}$$

where $[F]$ is the modal flexibility matrix; $[\phi]$ is the mass normalized modal vectors; and $[1/\omega^2]$ is a diagonal matrix containing the reciprocal of the squares of natural frequencies in ascending order. The modal contribution to the flexibility matrix decreases as the frequency increases, i.e., the flexibility matrix converges rapidly with increasing values of frequency. From only a few of the lower frequency modes, therefore, a good estimate of the flexibility can be made. The change in the flexibility matrix $\Delta[F]$ due to structural deterioration is given by

$$\Delta[F] = [F^d] - [F^h] \tag{2}$$

where the index h and d refer to the intact (or healthy) and the damaged states, respectively. Theoretically, structural deterioration reduces stiffness and increases flexibility. Increase in structural flexibility can, therefore, serve as a good indicator of the degree of structural deterioration.

Modal Strain Energy-based Damage Index

The strain energy U of a Bernoulli-Euler beam is given as follows:

$$U = \int \frac{EI}{2} \left(\frac{d^2 y}{dx^2} \right)^2 dx \tag{3}$$

where x is the distance measured along the length of the beam, y is the vertical deflection, EI is the flexural rigidity of the cross section and $d^2 y / dx^2$ is the curvature of the deformed beam.

Deterioration of a structure results in a reduction of its stiffness, which causes the changes in modal strain energy. The modal strain energy parameter is based on the relative differences in modal strain energy between an undamaged and damaged structure. The equation used to calculate the damage index β_{ji} for the j th element and i th mode of a beam is given below (Huth et al., 2005).

$$\beta_{ji} = \frac{\left(\int_{j}^{j+1} [\phi_i''^{*}(x)]^2 \, dx + \int_{0}^{L} [\phi_i''^{*}(x)]^2 \, dx \right) \int_{0}^{L} [\phi_i''(x)]^2 \, dx}{\left(\int_{j}^{j+1} [\phi_i''(x)]^2 \, dx + \int_{0}^{L} [\phi_i''(x)]^2 \, dx \right) \int_{0}^{L} [\phi_i''^{*}(x)]^2 \, dx} \tag{4}$$

In the above Eq., the terms $\phi_i''(x)$ and $\phi_i''^{*}(x)$ are the vectors of i-th mode shape curvature in undamaged and damaged structure, respectively, L is the length of the beam.

To account for all available modes, a single indicator for each location along the beam is given by

$$\beta_j = \frac{\sum_{i=1}^{NM} Num_{ji}}{\sum_{i=1}^{NM} Denom_{ji}} \tag{5}$$

where Num_{ji} = numerator of β_{ji} and $Denom_{ji}$ = denominator of β_{ji} in Eq. (4)

The strain energy of a plate of size $a \times b$ is given as follows:

$$U = \frac{D}{2} \int_{0}^{b} \int_{0}^{a} \left[\left(\frac{\partial^2 w}{\partial x^2} \right)^2 + \left(\frac{\partial^2 w}{\partial y^2} \right)^2 + 2v \left(\frac{\partial^2 w}{\partial x^2} \right) \left(\frac{\partial^2 w}{\partial y^2} \right) + 2(1-v) \left(\frac{\partial^2 w}{\partial x \partial y} \right)^2 \right] dxdy \tag{6}$$

where $D = Eh^3/12(1-v^2)$ is the bending stiffness of the plate, v is the Poisson's ratio, h is the plate thickness, w is the transverse displacement of the plate, x and y are coordinates along the plate edges, $\partial^2 w/\partial x^2$ and $\partial^2 w/\partial y^2$ are the bending curvatures, $2\partial^2 w/\partial x \partial y$ is the twisting curvature of the plate. For a particular mode shape $\phi_i(x, y)$, the strain energy U_i associated with that mode shapes is

$$U_i = \frac{D}{2} \int_0^b \int_0^a \left[(\frac{\partial^2 \phi_i}{\partial x^2})^2 + (\frac{\partial^2 \phi_i}{\partial y^2})^2 + 2v(\frac{\partial^2 \phi_i}{\partial x^2})(\frac{\partial^2 \phi_i}{\partial y^2}) + 2(1-v)(\frac{\partial^2 \phi_i}{\partial x \partial y})^2 \right] dxdy \quad (7)$$

If the plate is subdivided into N_x subdivisions in the x direction and N_y subdivisions in y the direction, then the energy U_{ijk} associated with sub-region jk for the i th mode is given by

$$U_{ijk} = \frac{D_{jk}}{2} \int_{b_k}^{b_{k+1}} \int_{a_j}^{a_{j+1}} \left[(\frac{\partial^2 \phi_i}{\partial x^2})^2 + (\frac{\partial^2 \phi_i}{\partial y^2})^2 + 2v(\frac{\partial^2 \phi_i}{\partial x^2})(\frac{\partial^2 \phi_i}{\partial y^2}) + 2(1-v)(\frac{\partial^2 \phi_i}{\partial x \partial y})^2 \right] dxdy \quad (8)$$

and hence

$$U_i = \sum_{k=1}^{N_y} \sum_{j=1}^{N_x} U_{ijk} \quad (9)$$

The fractional energy at location jk is defined to be

$$F_{ijk} = \frac{U_{ijk}}{U_i} \text{ and } \sum_{k=1}^{N_y} \sum_{j=1}^{N_x} F_{ijk} = 1 \quad (10, 11)$$

Similar expressions can be written using the modes ϕ_i^* of the damaged structure, where the superscript * indicates damaged state. A ratio of parameters can be determined that is indicative of the change of stiffness (or modal strain energy) in the structure as shown in Eqs. (12, 13).

$$\frac{D_{jk}}{D_{jk}^*} = \frac{f_{ijk}^*}{f_{ijk}} \quad (12)$$

where

$$f_{ijk} = \frac{\int_{b_k}^{b_{k+1}} \int_{a_j}^{a_{j+1}} \left[(\frac{\partial^2 \phi_i}{\partial x^2})^2 + (\frac{\partial^2 \phi_i}{\partial y^2})^2 + 2\nu(\frac{\partial^2 \phi_i}{\partial x^2})(\frac{\partial^2 \phi_i}{\partial y^2}) + 2(1-\nu)(\frac{\partial^2 \phi_i}{\partial x \partial y})^2 \right] dxdy}{\int_0^b \int_0^a \left[(\frac{\partial^2 \phi_i}{\partial x^2})^2 + (\frac{\partial^2 \phi_i}{\partial y^2})^2 + 2\nu(\frac{\partial^2 \phi_i}{\partial x^2})(\frac{\partial^2 \phi_i}{\partial y^2}) + 2(1-\nu)(\frac{\partial^2 \phi_i}{\partial x \partial y})^2 \right] dxdy} \quad (13)$$

and an analogous term f_{ijk}^* can be defined using the damaged mode shapes. In order to account for all measured modes, the following formulation for the damage index for sub-region jk is used:

$$\beta_{jk} = \frac{\sum_{i=1}^{m} f_{ijk}^*}{\sum_{i=1}^{m} f_{ijk}} \quad (14)$$

The complete derivation of the damage indicator for beam and plate are given in references (Stubbs et al., 1995; Cornwell et al., 1999), respectively.

APPLICATION

In order to demonstrate the practical application of the proposed multi-criteria procedure, two types of flexural members: (i) beams and (ii) plates are first developed as finite element (FE) models. Additional FE models with different damage scenarios are then developed for investigation. The natural frequencies and mode shapes of the first five modes of these models, before and after damage in the proposed damage scenarios, are extracted from the results of the free vibration FE analysis. These parameters can be used to determine the change of flexibility and the change of modal strain energy and thereby assess the structural health of the member concerned. The peak values of the two damage detection parameters indicate the corresponding simulated damage location. The accuracy of the damage detection method is then evaluated through observations of the plots of the two parameters, which are modal flexibility change and modal strain energy change. Detailed discussions on the finite element modelling of proposed structure are given below.

Case Studies 1: Finite Element Modeling and Analysis of Beams

Finite element models (FEM) of the undamaged and damaged simply supported beams are generated using the FE software SAP2000. Plane elements are used for the FE modelling. The details of the beam are given in Table 1. The flexural rigidity EI is assumed constant over the beam span, and damping effect is not taken into account. Modal analysis is performed to obtain the natural frequencies and the associated mode shapes of the beam. Further FE modelling and analysis are carried out on two-span and three-span continuous beam structures to extract the modal parameters. All continuous beams have the same span length of 2.8m and are simply supported at their ends, similar to the validated single-span beam model.

Table 1. Geometric And Material Properties Of Beam

Flexural member	Beam (2D)
Element type	Plane stress
Material	Steel
Length	2.8 m
Width	40 mm
Depth	20 mm
Poisson's ratio	0.3
Mass density	7850 kg/m^3
Modulus of elasticity	200 GPa

To simulate damage, the selected plane elements are removed from the bottom of the beams in the FE models. Nine such damage cases are investigated with two different sizes of flaws as listed in Table 2, in which size B flaw represents larger damage than size A flaw. The loss of second moment of area is 58% for all flaws. Parametric studies are carried out to investigate the feasibility of the multi-criteria damage detection approach on changes of parameters, such as damage severity and locations. Fig. 1 shows the first three damage scenarios in a single-span beam. In damage cases D1 and D2, a single damaged element is simulated on the beam at the mid-span with different damage severity to observe the changes of frequency, modal flexibility and modal strain energy-based damage index corresponding to the damage severity. The severity of damage increases in damage case D3 compared with D1 and D2, as two damaged elements are simulated on the beam, one located at the mid-span and the other at quarter-span. The other six damage scenarios in the two-span and three-span beams with different damage severity and locations are shown in Figs. 2 and 3, respectively. Fig. 4 shows the details of flaw size A simulated in the FE model.

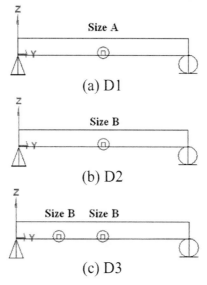

(a) D1

(b) D2

(c) D3

Figure 1. Damage case (D) for single-span beam (2.8m span length)

Table 2. Dimensions Of Flaws In Beam

Size	Length (mm)	Depth (mm)	Width (mm)
A	10	5	40
B	20	5	40

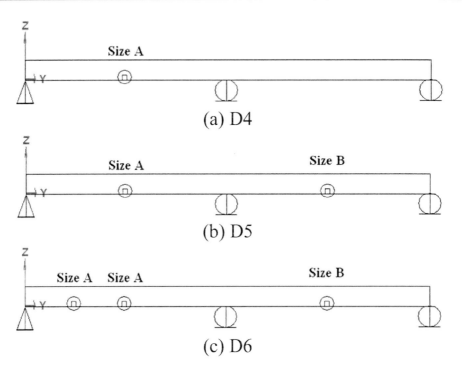

Figure 2. Damage case (D) for two-span beam (2.8m span length)

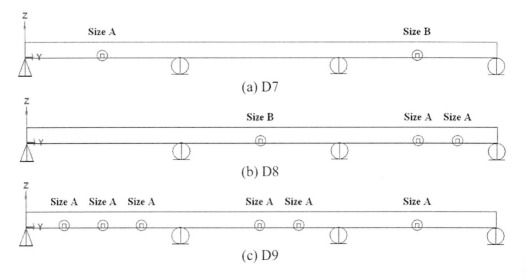

Figure 3. Damage case (D) for three-span beam (2.8m span length)

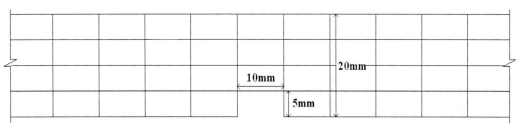

Figure 4. Flaw size A simulated in FEM

Frequency Change

The natural frequencies of the first five modes of the beam before and after damage in nine damage scenarios obtained from the results of the FEA are shown in Tables 3 and 4, respectively. Percentage changes in the natural frequencies between the undamaged and damaged states are listed within brackets. The first five vibration modes of undamaged FE model are plotted in Fig. 5. It is observed that the presence of damage in beams causes a decrease in the natural frequencies in all damage cases, with very few exceptions. If the damage cases D1 and D2 for the single-span beam are considered, change (i.e., decrease) in the frequency Δf is evident for the first, third and fifth modes, and there is no change for the second and fourth modes. This is because the damage elements are located at the nodes of these anti-symmetric modes of vibration and hence have no influence on the corresponding natural frequencies. By observing the changes in natural frequency of the first five modes, it is possible to achieve the Level 1 of identification that damage might be present in the beam-like structure when noise and environmental issues are not considered.

Table 3. Natural Frequencies Of Undamaged Beam From FEM

Member type	Boundary condition	Mode 1 f_1 (Hz)	Mode 2 f_2 (Hz)	Mode 3 f_3 (Hz)	Mode 4 f_4 (Hz)	Mode 5 f_5 (Hz)
	SS (1-span)	5.84	23.33	52.45	93.10	145.16
Beam	SS (2-span)	5.84	9.12	23.33	29.52	52.45
	SS (3-span)	5.84	7.48	10.92	23.33	26.58

Note: SS means simply supported.

Table 4. Natural Frequencies Of Damaged Beam From FEM (Percentage changes wrt the undamaged conditions are listed within brackets)

Damage case	Mode 1$_1$ (Hz)	Mode 2 f_2 (Hz)	Mode 3 f_3 (Hz)	Mode 4 f_4 (Hz)	Mode 5 f_5 (Hz)
D1	5.80 (0.68)	23.33 (0.00)	52.08 (0.71)	93.10 (0.00)	144.13 (0.71)
D2	5.77 (1.20)	23.33 (0.00)	51.90 (1.05)	93.10 (0.00)	143.65 (1.04)
D3	5.74 (1.71)	23.08 (1.07)	51.62 (1.58)	93.10 (0.00)	142.95 (1.52)
D4	5.82 (0.34)	9.10 (0.22)	23.33 (0.00)	29.51 (0.03)	52.26 (0.36)
D5	5.78 (1.03)	9.07 (0.55)	23.33 (0.00)	29.49 (0.10)	51.99 (0.88)
D6	5.77 (1.20)	9.05 (0.77)	23.25 (0.34)	29.39 (0.44)	51.90 (1.05)
D7	5.80 (0.68)	7.43 (0.67)	10.90 (0.18)	23.33 (0.00)	26.57 (0.04)
D8	5.79 (0.86)	7.45 (0.40)	10.86 (0.55)	23.28 (0.21)	26.48 (0.38)
D9	5.77 (1.20)	7.42 (0.80)	10.86 (0.55)	23.16 (0.73)	26.42 (0.60)

Note: Changes of natural frequencies result in decrease in all damage cases.

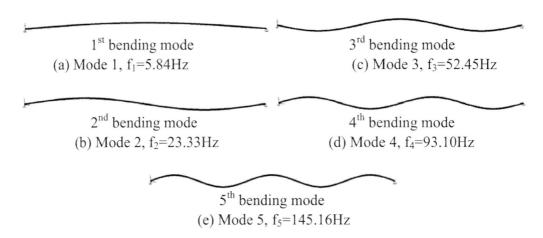

1st bending mode
(a) Mode 1, f_1=5.84Hz

3rd bending mode
(c) Mode 3, f_3=52.45Hz

2nd bending mode
(b) Mode 2, f_2=23.33Hz

4th bending mode
(d) Mode 4, f_4=93.10Hz

5th bending mode
(e) Mode 5, f_5=145.16Hz

Figure 5. First five vibration modes of single-span beam

Modal Flexibility Change

The first five natural frequencies and associated mode shapes obtained from the results of the FE analysis are used to calculate the modal flexibility change (MFC). The plots of MFC as a percentage along the beam for some representative cases are shown in Figs. 6(a)-(d). In all cases, the peak values correctly indicate the location of damage in the beams. Figs. 6(a) and (b) show the results for single-damage cases, and it is evident that the peak for the more severe damage case, D2, is higher than that for case D1. In Fig. 6(c), there are two unequal peaks corresponding to the two different damage locations in this beam, and once again, it is seen that greater damage in the beam is represented bigger peak in the MFC. Finally, Fig. 6(d) clearly shows that this damage case with triple damages has three distinct peaks in the MFC. The other damage cases showed analogous results and further confirm the feasibility of MFC in locating damage in a beam structure under a variety of damage scenarios. From the above observations, it is evident that the damage detection algorithm for MFC is able to locate the damage in the beam structure correctly in all damage cases and also give an indication of its severity. This confirms that the modal flexibility method is sufficiently sensitive to the damages in the beams.

Modal Strain Energy Change

The first five mode shapes and their corresponding mode shape curvatures obtained from the results of the FE analysis are used to calculate the modal strain energy-based damage index on beams. The plot of damage indices along the beam for damage cases D1, D2, D5 and D6 are shown in Figs. 6(e)-(h). The spikes with magnitudes greater than one indicate the location of damaged elements. Comparison of Figs. 6(e) and (f) shows that the peak in the damage index increases with the severity of damage. The peaks in Figs. 6(g) and (h) clearly indicate the multiple damages in the beam. From the results of all cases, it is evident that the damage index on the modal strain energy method is able to correctly locate the damage in beams in all damage cases.

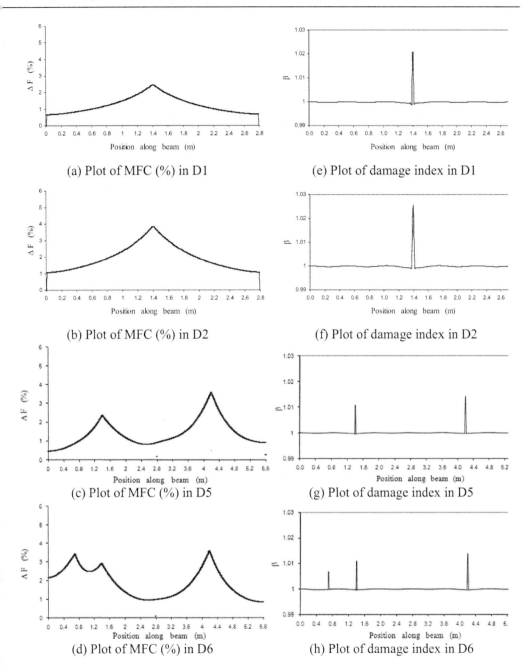

(a) Plot of MFC (%) in D1

(e) Plot of damage index in D1

(b) Plot of MFC (%) in D2

(f) Plot of damage index in D2

(c) Plot of MFC (%) in D5

(g) Plot of damage index in D5

(d) Plot of MFC (%) in D6

(h) Plot of damage index in D6

Figure 6. Modal flexibility change & Modal strain energy-based damage index on beam

Case Studies 2: Finite Element Modeling and Analysis of Plates

A rectangular steel plate with the size of 2.5m in length, 1m in width and 2mm in thickness is chosen for the numerical analysis. The material properties of the steel plate are listed in Table 5. The steel plate is divided into 250 plate elements. FE techniques are used to carry out modal analysis of the structure. Initially, a FE model of the steel plate clamped along all four edges is analyzed and the first three natural frequencies and the associated mode shapes of the plate obtained from the modal analysis are compared with those provided by Ulz and Semercigil (2008). The two sets of results are in good agreement as seen in Table 6, providing adequate confidence in the present FE modelling and analysis of plate structures. Additional FE models of single- and two-span plate structures are developed, and their modal analysis is carried out before and after damage. Damage is simulated by reducing the elastic modulus (E) to 50% and 80% in selected elements as shown in Figs. 7-9. An assumption is made that the mass of the plate does not change appreciably as a result of the damage. No structural damping is used in the FE analysis. Nine damage cases are investigated in this study. Among these cases, three different boundaries conditions in addition to different damage severities of selected elements are simulated to investigate the feasibility and capability of the multi-criteria damage detection approach. Fig. 7 shows three damage scenarios for the plate with all edges clamped. Figs. 8 and 9 show the other six damage scenarios for simply supported plates with single and two spans, respectively.

Table 5. Geometric And Material Properties Of Plate

Flexural member	Plate
Material	Steel
Length	2.5 m
Width	1 m
Depth	2 mm
Poisson's ratio	0.3
Mass density	7800 kg/m^3
Modulus of elasticity	210 GPa

Table 6. Validation Of FEM For Plate With Clamped Boundaries

Structural state	Frequency f_i	From Reference (Ulz and Semercigil, 2008) (Hz)	From SAP2000 (Hz)	Difference (%)
	f_1	11.81	11.78	0.3
Undamaged	f_2	13.89	13.77	0.8
	f_3	17.68	17.44	1.4

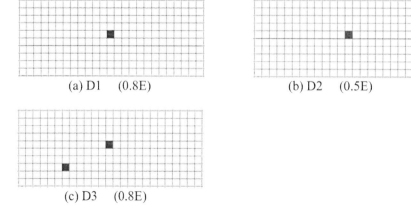

(a) D1   (0.8E)   (b) D2   (0.5E)

(c) D3   (0.8E)

Figure 7. Damage case (D) for plate with all edges clamped

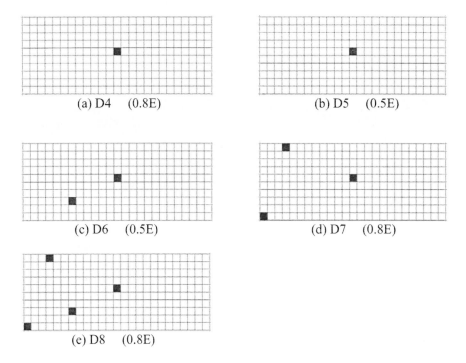

Figure 8. Damage case (D) for simply supported plate

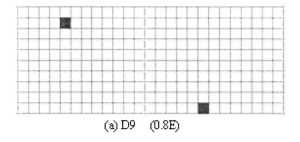

Figure 9. Damage case (D) for two-span plate

Frequency Change

The natural frequencies of the first five modes of the plates before and after damage obtained from the results of the FE analysis in nine damage scenarios are shown in Tables 7 and 8, respectively. Percentage changes in the natural frequencies between the undamaged and damaged conditions are listed within brackets. The first five vibration modes of undamaged FE model are plotted in Fig. 10. Again, it is observed that in general, the presence of damage in the plate causes a small decrease in the natural frequencies in all damage cases, with very few exceptions. If the damage cases D1 and D2 for the plate are considered, change (i.e., decrease) in the frequency Δf is evident for the first, third and fifth modes, while there is no change for the second and fourth modes. This is because the damage elements are located at the nodes of these anti-symmetric modes of vibration and hence have no influence on the corresponding natural frequencies. It may be concluded that by observing the changes in the natural frequencies, it is more possible to achieve Level 1 of identification of macro-damage in plate, rather than of micro-damage or small damage. The detection of small damage can be supplemented by advanced techniques such as acoustic emission monitoring.

Table 7. Natural Frequencies From FEM For Undamaged Plate

Member Type	Boundary condition (no. of span)	Mode 1 f_1 (Hz)	Mode 2 f_2 (Hz)	Mode 3 f_3 (Hz)	Mode 4 f_4 (Hz)	Mode 5 f_5 (Hz)
	Edges clamped (1-span)	11.78	13.77	17.44	22.91	30.15
Plate	Simply supported (1-span)	0.76	2.66	3.07	5.95	6.95
	Simply supported (2-span)	3.04	4.90	5.88	7.36	12.19

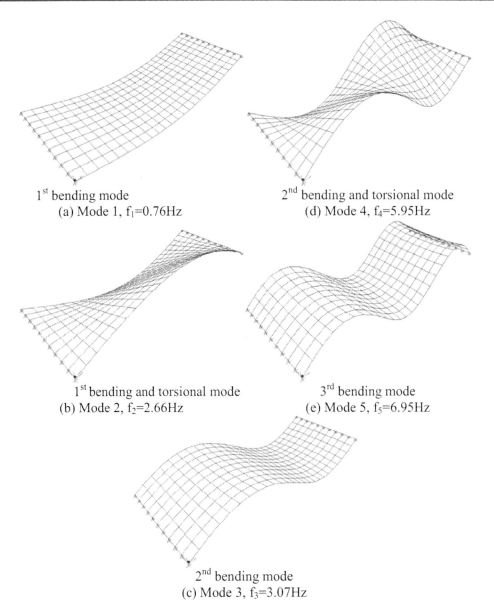

1st bending mode
(a) Mode 1, f_1=0.76Hz

2nd bending and torsional mode
(d) Mode 4, f_4=5.95Hz

1st bending and torsional mode
(b) Mode 2, f_2=2.66Hz

3rd bending mode
(e) Mode 5, f_5=6.95Hz

2nd bending mode
(c) Mode 3, f_3=3.07Hz

Figure 10. First five vibration modes of undamaged plate with simply supported condition

Table 8. Natural Frequencies From FEM For Damaged Plate

Damage case	Mode 1 f_1 (Hz)	Mode 2 f_2 (Hz)	Mode 3 f_3 (Hz)	Mode 4 f_4 (Hz)	Mode 5 f_5 (Hz)
D1	11.77 (0.14)	13.77 (0.01)	17.42 (0.12)	22.91 (0.00)	30.11 (0.15)
D2	11.74 (0.39)	13.77 (0.01)	17.38 (0.38)	22.91 (0.02)	30.02 (0.43)
D3	11.77 (0.14)	13.77 (0.02)	17.41 (0.15)	22.90 (0.04)	30.10 (0.18)
D4	0.76 (0.08)	2.66 (0.00)	3.07 (0.00)	5.94 (0.06)	6.95 (0.08)
D5	0.76 (0.24)	2.66 (0.00)	3.07 (0.00)	5.94 (0.17)	6.94 (0.23)
D6	0.76 (0.13)	2.66 (0.04)	3.06 (0.08)	5.94 (0.09)	6.95 (0.11)
D7	0.76 (0.11)	2.66 (0.15)	3.07 (0.07)	5.93 (0.19)	6.94 (0.18)
D8	0.76 (0.13)	2.66 (0.04)	3.06 (0.08)	5.94 (0.09)	6.95 (0.11)
D9	3.03 (0.16)	4.90 (0.12)	5.87 (0.10)	7.36 (0.10)	12.18 (0.08)

Modal Flexibility Change

Plots of MFC for damage cases D1, D2 and D9 are shown in Figs. 11(a), (b) and (d), respectively. To better demonstrate the damage detection results for the plate structure, the plot of MFC for damage case D8, as shown in Fig. 11(c), is expressed as a percentage with respect to the undamaged modal flexibility matrix. The peak values indicate the location of damage in the plate. Comparison of Figs. 11(a) and (b) pertaining to single damage in a plate shows that as the severity of the single damage at mid-span increases, the corresponding MFC also increases, as demonstrated by the higher peak. For the case of multiple-damage detection in damage case D9, the modal flexibility method is able to correctly locate the damage. The other damage cases showed analogous results and further confirm the feasibility of MFC in locating damage in a beam structure under a variety of damage scenarios. However for multiple-damage case D8, the results in Fig. 11(c) could not clearly indicate the damage locations, and the damage indicator has missed the damage at the mid-span of the plate. Overall, the results show that the modal flexibility method is able to correctly locate the damage in most multiple-damage cases, except in cases D7 and D8, where the damage indicator seems to have missed the damage at the mid-span of the plate. Similar to cases where damage at nodes of vibrating modes could not affect the corresponding natural frequencies, this problem in plates is a further evidence for the need of multi-criteria damage assessment.

Modal Strain Energy Change

The modal strain energy-based damage index for plate is calculated by using Eqs. (13) and (14). The plot of damage indices for damage cases D1, D2, D8 and D9 are shown in Figs. 11(e)-(h), respectively. The spikes with the magnitudes greater than one indicate the location of damaged elements. The peak in Fig. 11(f) corresponding to a more severe damage case is higher than the peak in Fig. 11(e). Multi-peaks in Figs. 11(g) and (h) indicate the locations of multiple damages correctly in the plate. Overall, the results show that the strain energy method is capable of detecting multiple damages in plates for all damage cases.

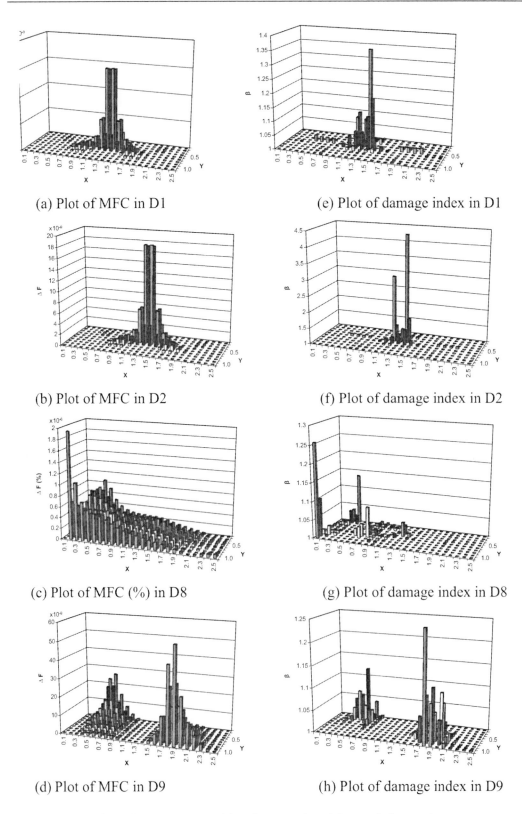

(a) Plot of MFC in D1

(e) Plot of damage index in D1

(b) Plot of MFC in D2

(f) Plot of damage index in D2

(c) Plot of MFC (%) in D8

(g) Plot of damage index in D8

(d) Plot of MFC in D9

(h) Plot of damage index in D9

Figure 11. Modal flexibility change & Modal strain energy-based damage index on plate

CONCLUSION

Dynamic computer simulation techniques are used to develop and apply multi-criteria-based non-destructive damage detection methodology for beams and plates. The proposed procedure involves two vibration-based damage detection parameters (i) changes in modal flexibility matrix and (ii) changes in modal strain energy-based damage index, in addition to changes in natural frequencies, all of which are evaluated from the results of free vibration analysis of the undamaged and damaged finite element models of the structure. Based on the results from the extensive dynamic computer simulations, it is found that the proposed multi-criteria approach is feasible for damage assessment in the flexural members. As a starting point, changes in natural frequencies can be used to detect the presence of damage, since this can be done from a single point measurement. This approach could provide an inexpensive structural assessment technique as frequency measurement is easily acquired. The presence of damage could be triggered by a change in natural frequency using high-sensitivity transducers. The modal flexibility and modal strain energy methods can then be used to locate the damage in the proposed structures. The most attractive feature of the two methods is that they can be implemented using the first few vibration modes. The changes in the modal flexibility matrix and modal strain energy between the undamaged and damaged structure provide a basis for locating the damage. Overall, it can be concluded that the multi-criteria vibration-based approach provides more reliable and accurate tools for identification of multiple damages. As there are some discrepancies in both damage assessment parameters (MFC and MSEC), the multi-criteria approach should be able to provide more accurate damage assessment (Shih et al., 2009a, 2009b). Due to the major advances in the fields of structural dynamics and experimental modal analysis, the multiple criteria approach shows promise in being used for detecting and locating damages in structures.

REFERENCES

Araújo dos Santos, J.V., Maia, N.M.M., Mota Soares, C.M. and Mota Soares, C.A. (2008). "Structural damage identification: a survey." In: Topping BHV, Papadrakakis M, editors. *Trends in Computational Structures Technology*, Saxe-Coburg Publications, Stirlingshire, UK, Chapter 1, 1-24. doi:10.4203/csets.19.1.

Baghiee, N., Esfahani, M.R. and Moslem, K. (2009). "Studies on damage and FRP strengthening of reinforced concrete beams by vibration monitoring." *Engineering Structures*, 31(4), 875-893.

Choi, F. C., Li, J., Samali, B. and Crews, K. (2008). "Application of the Modified Damage Index Method to Timber Beams." *Engineering Structures*. 30(4) pp. 1124-1145.

Cornwell, P., Doebling, S.W. and Farrar, C.R. (1999). "Application of the strain energy damage detection method to plate-like structures." *Journal of Sound and Vibration*, 224(2), 359-374.

Curadelli, R.O., Riera, J.D., Ambrosini, D. and Amani, M.G. (2008). "Damage detection by means of structural damping identification." *Engineering Structures*, 30(12), 3497-3504.

Doebling, S.W., Farrar, C.R., Prime, M.B. and Shevitz, D.W. (1996). "Damage identification and health monitoring of structural and mechanical systems from changes in their vibration characteristics: a literature review." *Report no. LA-13070-MS*, Los Alamos National Laboratory, Los Alamos (USA).

El-Ouafi Bahlous, S., Smaoui, H. and El-Borgi, S. (2009). "Experimental validation of an ambient vibration-based multiple damage identification method using statistical modal filtering." *Journal of Sound and Vibration*, 325(1-2), 49-68.

Fang, S.E. and Perera, R. (2009). "Power mode shapes for early damage detection in linear structures." *Journal of Sound and Vibration*, 324(1-2), 40-56.

Farrar, C.R. and Cone, K.M. (1995). "Vibration testing of the I-40 Bridge before and after the introduction of damage." *Proceedings of 13th IMAC*, Vol.I, 203-209.

Fernandes, K.M., Stutz, L.T., Tenenbaum, R.A. and Silva Neto, A.J. (2008). "Vibration and wave propagation approaches applied to assess damage influence on the behavior of Euler-Bernoulli beams: Part I direct problem." In: Topping BHV, Papadrakakis M, editors. *Proceedings of the Ninth International Conference on Computational Structures Technology*, Civil-Comp Press, Stirlingshire, UK, Paper 38. doi:10.4203/ccp.88.38.

Ge, M. and Lui, M. (2005). "Structural damage identification using system dynamic properties." *Computers and Structures*, 83(27), 2185-2196.

Guan, H. and Karbhari, V.M. (2008). "Improved damage detection method based on Element Modal Strain Damage Index using sparse measurement." *Journal of Sound and Vibration*, 309(3-5), 465-494.

Housner, G.W., Bergman, L.A., Caughey, T.K., Chassiakos, A.G., Claus, R.O., Masri, S.F., Skelton, R.E., Soong, T.T., Spencer, B.F. and Yao, J.T.P. (1997). "Structural control: past, present, and future." *Journal of Engineering Mechanics*, 123(9), 897-971.

Huth, O., Maeck, J., Kilic, N. and Motavalli, M. (2005). "Damage identification using modal data: experiences on a prestressed concrete bridge." *Journal of Structural Engineering*, 131(12), 1898-1910.

Lee, J.J. and Yun, C.B. (2006). "Damage diagnosis of steel girder bridges using ambient vibration data." *Engineering Structures*, 28(6), 912-925.

Lee, L.S., Karbhari, V.M. and Sikorsky, C. (2004). "Investigation of integrity and effectiveness of RC bridge deck rehabilitation with CFRP composites." *SSRP-2004/08*, Department of Structural Engineering, University of California, San Diego.

Peng, Z.K., Lang, Z.Q. and Chu, F.L. (2008). "Numerical analysis of cracked beams using nonlinear output frequency response functions." *Computers & Structures*, 86(17-18), 1809-1818.

Perera, R., Ruiz, A. and Manzano, C. (2007). "An evolutionary multiobjective framework for structural damage localization and quantification." *Engineering Structures*, 29(10), 2540-2550.

Petryna, Y.S., Ahrens, A. and Stangenberg, F. (2006). "Damage simulation and health assessment of a road bridge." In: Topping BHV, Montero G, Montenegro R, editors. *Proceedings of the Eighth International Conference on Computational Structures Technology*, Civil-Comp Press, Stirlingshire, UK, Paper 2. doi:10.4203/ccp.83.2.

Petryna, Y.S., Krätzig, W.B. and Stangenberg, F. (2002). "Structural damage: simulation and assessment." In: Topping BHV, Bittnar Z, editors. *Computational Structures Technology*, Saxe-Coburg Publications, Stirlingshire, UK, Chapter 14, 351-377. doi:10.4203/csets.7.14.

Ren, W.X. and Sun, Z.S. (2008). "Structural damage identification by using wavelet entropy." *Engineering Structures*, 30(10), 2840-2849.

Rytter, A. (1993). "Vibration-based inspection of civil engineering structures." *Doctoral Dissertation*, Department of Building Technology and Structural Engineering, University of Aalborg, Aalborg, Denmark.

Shih, H.W., Thambiratnam, D.P. and Chan, T.H.T. (2009a). "Vibration-based structural damage detection in flexural members using multi-criteria approach." *Journal of Sound & Vibration,* 323(3-5), 645–661.

Shih, H.W., Thambiratnam, D.P. and Chan, T.H.T. (2009b). "Damage assessment in structures using vibration characteristics." *Doctoral Dissertation*, School of Urban Development, Queensland University of Technology, Brisbane, Australia.

Stubbs, N., Kim, J.T. and Farrar, C.R. (1995). "Field verification of a non-destructive damage localization and severity algorithm." *13th International Modal Analysis Conference*, 210-218.

Ulz, M. H., and Semercigil, S. E. (2008). "Vibration control for plate-like structures using strategic cut-outs." *Journal of Sound and Vibration,* 309(1-2), 246-261.

Xia, Y., Hao, H., and Deeks, A.J. (2007). "Dynamic assessment of shear connectors in slab-girder bridges." *Engineering Structures*, 29(7), 1475-1486.

Zacharias, K., Douka, E., Hadjileontiadis, L.J. and Trochidis, A. (2008). "Non-linear vibration technique for crack detection in beam structures using frequency mixing." In: Topping BHV, Papadrakakis M, editors. *Proceedings of the Ninth International Conference on Computational Structures Technology*, Civil-Comp Press, Stirlingshire, UK, Paper 123. doi:10.4203/ccp.88.123.

Zapico, J.L. and Gonzalez, M.P. (2006). "Vibration numerical simulation of a method for seismic damage identification in buildings." *Engineering Structures,* 28(2), 255-263.

In: Structural Health Monitoring in Australia   ISBN: 978-1-61728-860-9
Editors: Tommy H.T. Chan and D. P. Thambiratnam   ©2011 Nova Science Publishers, Inc.

Chapter 3

USE OF ACOUSTIC EMISSION TECHNIQUE FOR STRUCTURAL HEALTH MONITORING OF BRIDGES

Manindra Kaphle*, Andy C.C. Tan‡,
David P. Thambiratnam≠ and Tommy H.T. Chan±

ABSTRACT

Bridges are valuable assets of every nation. They deteriorate with age and often are subjected to additional loads or different load patterns than they were originally designed for. These changes in loads can cause localized distress and may result in bridge failure if not corrected in time. Early detection of damage and appropriate retrofitting will aid in preventing bridge failures. Large amounts of money are spent in bridge maintenance all around the world. Hence, a need exists for a reliable, cost-effective technology capable of monitoring the structural health of bridges, thereby ensuring they operate safely and efficiently during the whole intended lives. Monitoring of bridges has been traditionally done by means of visual inspection. Visual inspection alone is not capable of locating and identifying all signs of damage, hence a variety of structural health monitoring (SHM) techniques is used regularly nowadays to monitor performance and to assess condition of bridges for early damage detection. Acoustic emission (AE) is one such technique that is finding an increasing use in SHM applications of bridges all around the world. The chapter presents a brief introduction to structural health monitoring and techniques commonly used for monitoring purposes. Theory behind the acoustic emission technique, wave nature of AE, some previous applications and remaining challenges in its use as a SHM technique are also discussed. Scope of the project currently undertaken and work carried out so far will be explained, followed by some recommendations for the work planned in the future.

---

* BEE, QUT, GPO Box, 2434 Q4001 (manindra.kaphle@student.qut.edu.au)
‡ BEE, QUT, GPO Box, 2434 Q4001 (a.tan@qut.edu.au)
≠ BEE, QUT, GPO Box, 2434 Q4001 (d.thambiratnam@qut.edu.au)
± BEE, QUT, GPO Box, 2434 Q4001 (tommy.chan.@qut.edu.au)

INTRODUCTION

Structural health monitoring (SHM) is a field that has been receiving a considerable amount of attention in various areas of mechanical, civil and aerospace engineering. Researchers in the area have defined SHM in various ways. Achenbach (2009) defined SHM as a system that provides continuous or on-demand information about the state of a structure, so that an assessment of the structural integrity can be made at any time, and timely remedial actions may be taken as necessary. Similarly, Sohn et al. (2003) explained the process of SHM as observing a system over time using periodically sampled dynamic response measurements from an array of sensors, the extraction of damage-sensitive features from these measurements, and the statistical analysis of these features to determine the current state of the system's health. A range of techniques is regularly used to monitor the health of engineering structures such as aircraft, pipelines and nuclear reactors. Nowadays, civil infrastructure such as bridges and buildings are also being monitored in order to ascertain their structural integrity and to predict any damage early.

Bridges are valuable assets of every nation. Many bridges in use today were built decades ago and are now subjected to added loads due to increased traffic loads or changes in load patterns than originally designed for. These load conditions and deterioration with age can cause localized distress and may even result in bridge failure, if not corrected in time. Bridge failures, though rare, can cause huge financial losses as well as loss of lives. A recent example is the I-35W highway bridge collapse in Minnesota, USA, in August 2007, which resulted in 13 deaths and 145 injuries. In USA, out of a total 593,416 bridges, 158,182, that is, around 26.7 percent, were identified as being either structurally deficient or functionally obsolete (USDoT, 2006). Large sums of money are spent in bridge maintenance all around the world. In Australia alone, annual maintenance expenditure on about 33,500 total number of bridges runs around 100 million dollars (Austroads, 2004). All these statistics point to the need of a reliable technology capable of monitoring the structural health of bridges, thereby ensuring safe and efficient operation during the whole intended lives. Such technology will also offer economic benefits to the bridge owner in scheduling maintenance and replacement works.

SHM Techniques

Visual inspection has been the traditional means of monitoring bridges. Trained personnel inspect bridges in regular intervals to check the presence of any signs of damage and recommend appropriate retrofitting if necessary. This is a simple method, and use of dye penetrant can facilitate the inspection process. But small or hidden cracks are hard to locate, and cracks due to corrosion or fatigue may go undetected until they reach critical stage (Holford et al., 2001). Therefore, more reliable methods are often needed. A wide array of methods is routinely used for structural health monitoring of bridges. These methods can be broadly classified as global and local methods. Vibration-based monitoring techniques usually give a global picture, indicating the presence of damage in the entire structure. These are based on the principle that the changes in the global properties (mass, stiffness and damping) of a structure cause a change in its modal properties (such as natural frequencies and mode shapes). The modal properties or the quantities derived from them such as modal flexibility and modal strain energy can then be used for damage identification (Chang et al., 2003; Farrar et al., 2001; Shih et al., 2009). These global methods are common in use, but main drawback is that due to the large size of bridges, some damage may only cause negligible change in dynamic properties and thus may go unnoticed. Moreover, in order to find the exact location of damage, local methods are often better alternatives.

Several non-destructive techniques (NDT) are available for local structural health monitoring, and most commonly used ones in bridge health monitoring are based on the use of fibre optics (though they can also be used for global monitoring), electromagnetic waves (Magnetic particle testing, Eddy current testing and radiographic techniques) and mechanical waves (ultrasonic and acoustic emission techniques). A brief comparison of these NDT methods is given in Table 1.

Table 1. Comparison of different NDT methods

NDT method	Principle	Pros	Cons
Fibre optics	– Capable of sensing a variety of perturbations, mainly used to sense strain and temperature – The three sensing mechanisms of optical fibres are based on intensity, wavelength, and interference of the lightwave (Ansari, 2007)	– Geometric conformity – No electric interference – Can be used for a wide range of civil structures such as buildings, bridges, pipelines, (Li et al., 2004)	– Costly – Need highly trained professionals
Magnetic particle testing	– Use powder to detect leaks of magnetic flux (Rens et al., 1997)	– Economic	– Not applicable for nonferrous materials
Eddy current testing	– Presence of a flaw changes the eddy- current pattern (Chang and Liu, 2003)	– Can detect crack through paint – Effective for detecting cracks in welded joints	– Expensive and can be used only for conducting materials – Sensor mounting can be troublesome
Radiographic	– Radiographic energy source generates radiation and is captured by recording medium in other side of specimen (Chang and Liu, 2003).	– Promising laboratory results	– Large size of equipment – Health hazard
Ultrasonic	– Transducers are used to introduce high frequency waves into a specimen and receive the pulses – Inhomogeneities in the material induce changes to the propagating waves (Mancini, et al. 2006)	– Position of flaw can be determined	– Expensive – Coupling of sensor with the specimen surface may create problem – Requires generation of source signal
Acoustic emission (AE)	– AE waves are elastic stress waves that arise from the rapid release of energy inside material, for example from crack initiation (Carlos, 2003). – AE technique involves recording the waves by sensors and then analyzing the signals to extract information about the source of emission.	– Highly sensitive – Ability to easily locate damage that acts as emission source – Passive technique, no energy need to be supplied (unlike ultrasonic method)	– Background noises affect monitoring in large structures – High sampling rates generate large volumes of data

ACOUSTIC EMISSION TECHNIQUE

As explained in Table 1, acoustic emission (AE) waves are stress waves that arise from the rapid release of energy from localized sources within a material (Carlos, 2003). Some common sources of AE in engineering materials include initiation and growth of cracks, yielding, failure of bonds, fibre failure and pullout in composites. A diagrammatic representation of acoustic emission technique can be seen in **Figure 1**, where under the application of stress, a crack originates in the specimen and acts as a source of AE waves. These waves propagate in all directions and can be recorded by a sensor placed on the surface. The signals are amplified and sent to the AE acquisition system for filtering and further processing, which will then provide vital information about the nature of the source.

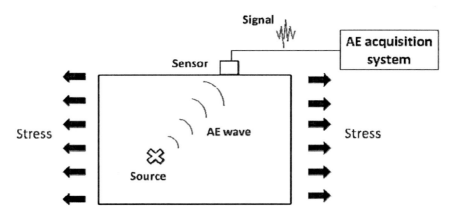

Figure 1. Acoustic emission technique

A typical AE signal with some associated signal parameters can be seen in Figure 2.

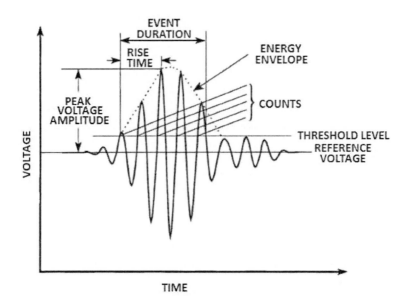

Figure 2. AE signal parameters (Lozev et al., 1997)

An event or hit occurs when a signal exceeds the set threshold value and triggers the acquisition system to record data. The threshold value is set to remove as much noise as possible, but care should be taken that no weak signals are missed by setting too high threshold value. Rise time is the interval between the time a signal is triggered and the time the signal reaches the maximum amplitude. Similarly, duration is the interval between the time a signal is triggered and the time the signal decreases below the threshold value. Counts are the number of times a signal crosses the threshold with the duration. Counts rate and events rate are also used regularly. Peak amplitude and energy of the signal are parameters that convey information about the strength of the AE source. Various ways of expressing energy exist, such as area under the amplitude curve and RMS (root mean square) value.

For analysis of recorded AE data, signal parameters discussed above can be used. In such parameter-based approach, the signal itself is not recorded but only some parameters are recorded and used for analysis purposes. This minimizes the amount of data stored and enables fast data recording. But with the availability of better sensors and higher computing resources, it is now possible to perform quick data acquisition and record the whole waveform (Huang et al., 1998). This waveform-based approach offers better data interpretation capability than parameter-based approach by allowing the use of signal-processing techniques and in aiding in signal-noise discrimination (Grosse and Linzer, 2008). Furthermore, it is believed that though waveform shape is affected by the nature and geometry of the medium of propagation, it still contains information about the source mechanisms. Hence, analysis of waveforms can be expected to provide information about the nature of the source and help in distinguishing different sources of AE. AE waves can travel in different modes and waveform-based analysis is necessary to understand the modes of travel. The main disadvantage of waveform-based approach is the generation of large volume of data, but use of signal-processing tools can help in data conditioning, thereby minimizing the amount of data that has to be stored.

Modal Nature of AE Waves

AE waves are elastic stress waves and travel in solids in various modes. Four main types of AE waves can be identified as: longitudinal waves, transverse waves, surface waves and Lamb (plate) waves. Reflected waves and diffracted waves are also usually present. These waves are governed by the same set of partial differential wave equation but require satisfaction of different sets of physical boundary conditions (Rose, 1999). A general overview of the AE wave modes is given next.

Longitudinal and shear waves
Both are collectively known as body or bulk waves. Longitudinal waves are also known as compression, primary or P waves. In longitudinal waves, particles oscillate in the direction of wave propagation. Transverse waves are also known as shear or S waves, and the oscillations occur transverse to the direction of propagation. Mode conversions can occur between P waves and S waves. Diagrammatic representation of longitudinal and transverse wave modes can be seen in **Figure 3** and **Figure 4**, respectively.

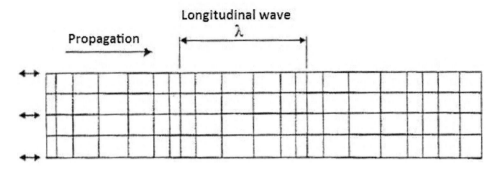

Figure 3. Longitudinal waves (Holford and Lark, 2005)

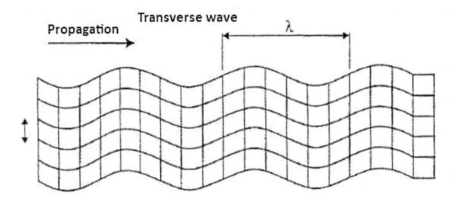

Figure 4. Transverse waves (Holford and Lark, 2005)

Surface Waves

Surface waves, also known as Rayleigh waves, travel on the surface of semi-infinite solid. They arise due to the interaction of longitudinal and shear waves on the surface and travel with velocity slightly slower than that of shear waves. Figure 5 shows the propagation of surface waves.

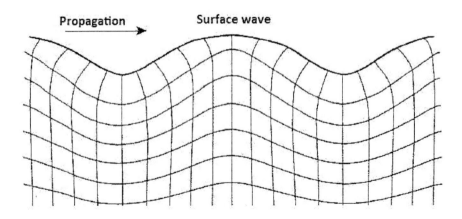

Figure 5. Surface waves (Holford and Lark, 2005)

Lamb Waves

Lamb waves are common in plate-like structures. They consist of two basic modes: an extensional or symmetric (S_0) mode that often appears as higher velocity but lower amplitude waves preceding flexural or asymmetric (A_0) mode (Finlayson et al., 2003; Maji et al., 1997). Higher order modes (S_1, A_1, S_2, A_2) can occur as well. The velocities of the Lamb wave modes depend on thickness of the plate and frequencies. Dispersion curves, which are based on the solution of Lamb wave equations, show the variation of the modes with the product of plate thickness and frequency (Holford and Lark, 2005). Propagation of basic lamb wave modes is illustrated in Figure 6 and Figure 7.

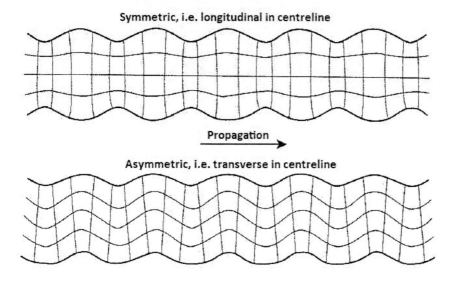

Figure 6. Lamb waves: symmetric (extensional) and asymmetric (flexural) modes (Holford and Lark, 2005)

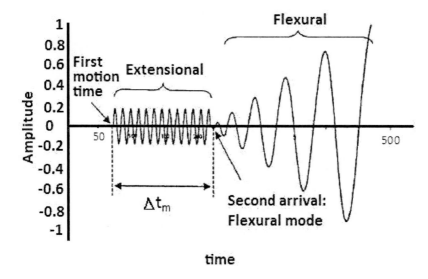

Figure 7. Theoretical symmetric and asymmetric modes (Finlayson et al. ,2003)

Previous Applications of AE for Bridge Monitoring

Gong et al. (1992) used AE technique to detect fatigue crack initiation and to monitor fatigue crack growth on steel railroad bridges. Inspection records, engineering judgements, results from FEA or strain gauge were mentioned as suitable criteria that can be used to select monitoring sites and weld repairs, edges of bolt and rivet holes and connected areas were identified as locations mainly likely to experience fatigue crack. Using AE count rate, cracks were grouped into different levels of severity. AE technique was concluded to be successful in finding new cracks, in identifying active cracks and in validating the effectiveness of repairs; but noises such as rubbing and fretting of bridge parts were identified as troublesome as they generated high frequency signals similar to crack growth signals.

In report submitted to the Virginia Transportation Research Council, Lozev et al. (1997) presented the results of field testings in five bridges as well as laboratory fatigue tests carried out to characterize acoustic emission associated with steel cracking and various sources of noise in a typical bridge environment. Monitoring was done on the regions where cracks had been discovered, such as pin and hanger connections with crack, cracked web of girder that had been retrofitted and cracked welded region. Source location, strain data and frequency analysis of waveform were used to discriminate sources. AE technique was found suitable in distinguishing active and benign cracks, but presence of spurious noise sources was identified as the main impediment to the successful use of AE in bridge inspection. It was found uniqueness of waveforms could be used to distinguish different sources, but manual classification was hard when a large number of waveforms were present.

In a study sponsored by the Minnesota Department of Transportation, Mckeefry and Shield (1999) studied fatigue crack propagation and checked for effectiveness of retrofits in repairing crack growth in steel bridge girders. Both laboratory study and field implementation were carried out. Strain gauges were used to study stress distribution. AE monitoring was found successful in detecting propagating and extinguished cracks. It was concluded that inspection of the retrofit periodically was necessary to check for continued crack propagation.

Yoon et al. (2000) performed tests in corroded reinforced concrete specimens to identify different sources of damage, such as micro crack development, localized crack propagation and debonding of the reinforcing steel. Analysis of both AE parameters and waveform was seen to be useful in estimating the damage of reinforced concrete structures.

Rizzo and di Scalea (2001) performed tests on carbon-fibre-reinforced-polymer bridge stay cables by recording the counts, amplitudes and energy of the signals. Amplitude values and frequency components of the AE were found to provide a qualitative correlation with the type of occurred damage. Need to give wave attenuation and dispersion a proper care was highlighted as even large signals may not be recorded if present far away from sensors.

Melbourne and Tomor (2006) studied crack propagation and failure mechanisms in masonry arches under static and long-term cyclic loading using AE technology and found it effective in locating damaged regions and predicting potential areas of failure.

Yuyama et al. (2007) analyzed high-strength tendon of prestressed concrete bridges to detect and locate corrosion-induced failure and found that analysis of detected AE signal parameters such as amplitude and signal duration allowed distinguishing meaningful AE events from other sources as traffic noises.

Summary Of Previous Applications

Thus, it is clear that AE technique has been attempted for early damage detection and health assessment of different types of bridge structures. Most of the studies have combined experimental testing in laboratory with field testing in real bridges. Most uses have been for local monitoring—the areas where crack presence is known are monitored using a number of sensors. The aims are primarily to determine if the cracks are active or not (by recording AE events) and to check how effective the retrofits are. Fatigue cracking has been identified as serious problem especially in steel structures, and load cycles are applied in laboratory tests to initiate fatigue cracking. Passing vehicles act as the sources of loading in real-life testing. Experimental studies have generally given encouraging results, whereas real-life tests have suffered mainly from the presence of other sources that give rise to signals that can mask the signals from cracks; therefore, source discrimination has been identified as an important task. Parameter-based approach and waveform-based approach have been used to determine the onset of damage, quantifying damage and distinguishing sources of AE. Waveform-based approach is prevalent in newer studies and is preferred as it provides frequency information of the signals and allows the use of signal-processing tools for removing noises. Use of additional information, such as strain, is often made to relate AE activity with load cycles. Main challenges encountered will be discussed in next section.

Challenges

Monitoring bridges presents practical challenges, as bridges are usually large in size, may have complex shapes and are composed of different materials. Gaining access to all areas of a bridge may also be difficult. In addition to general practical issues, other challenges exist in using AE technique for SHM of bridge structures. Some of the commonly encountered challenges and the ways suggested to address them are discussed next.

Sensor Placement and Selection

Selection and optimal placement of sensors are both important for detecting damage. As AE waves propagate in a material their amplitude decreases, and this is known as attenuation. Reasons include dispersion, scattering and conversion to other forms of energy such as heat. Material properties affect attenuation; for example, waves attenuate faster in concrete compared to steel. Because of attenuation, waves can be recorded only up to a certain distance and this places a limit on the distance of separation of the sensors. The number of sensors can also be limited by the available channels in signal analyzing systems, access to bridge locations or because of economic reasons. Therefore, careful selection of regions of structures where flaws are likely to occur is necessary for sensor placement. Good coupling is also necessary between the sensors and the test specimen to ensure waves are transmitted properly.

Sensors are of two types: broad-band and resonant. Broad-band sensors have low sensitivity and may record additional background noise, while resonant sensor of certain frequency may miss important signals (Golaski et al., 2002). Resonant sensors tend to resonate at their characteristic frequency, regardless of the source, thus distorting the original wave (Gorman, 1998). Sison et al. (1998) recommended the use of wideband AE sensors rather than resonant sensors for bridge testing, as the former are more capable of distinguishing between different AE sources. A solution is to use broad-band sensor for initial tests in sample specimen and using this frequency response to select resonant sensor.

Data Management

Brownjohn et al. (2005) identified data management and storage, wireless data transmission, data mining, evaluating performance against structural models, and presentation of minimal and reliable information to bridge managers for decision making as one of the main challenges for health monitoring of civil structures. AE data is collected at high sampling rates, often in the order of 1 MHz or more, so testing can generate a large amount of data. Though all AE data is not useful as most acoustic emission arises from spurious sources and hence only selected data need to be stored, continuous monitoring of a structure for a long period of time can still generate huge amount of data, and effective data management is crucial to establish an efficient AE-based system. Wireless sensing of structures offers added benefits and new techniques are also being investigated (Grosse et al., 2006).

Source Localization

Finding the location of the source of damage is an important part of the monitoring process. Time of arrival (TOA) method based on first threshold crossing is the most common way of determining the source location of acoustic emission. An illustration of two-dimensional source location can be seen in Figure 8.

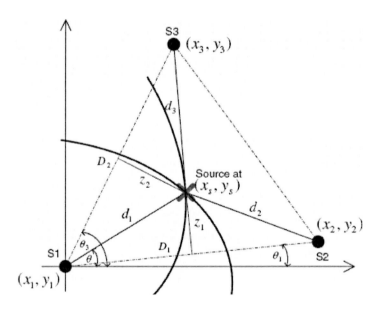

Figure 8. Two-dimensional source location (Nivesrangsan et al., 2007)

Three sensors S1, S2 and S3 are placed on the surface of a structure at locations (x_1, y_1), (x_2, y_2) and (x_3, y_3). The location of the source, to be determined, is (x_s, y_s). The distances between the sensors are D_1, D_2 and D_3, and the distances between the source and the sensors are d_1, d_2 and d_3 as marked.

Using the derivations made by Nivesrangsan et al. (2007), the distance d_1 can be written as:

$$d_1 = \frac{D_1^2 - \Delta t_1^2 \cdot c^2}{2(\Delta t_1 \cdot c + D_1 \cos(\theta - \theta_1))} \tag{1}$$

$$d_1 = \frac{D_2^2 - \Delta t_2^2 \cdot c^2}{2(\Delta t_2 \cdot c + D_2 \cos(\theta_3 - \theta))} \tag{2}$$

where angles θ, θ_1 and θ_3 are marked in Figure 8 and

$$d_2 - d_1 = c(t_2 - t_1) = \Delta t_1 \cdot c \tag{3}$$

$$d_3 - d_1 = c(t_3 - t_1) = \Delta t_2 \cdot c \tag{4}$$

t_1, t_2 and t_3 are the times the signal reaches sensors S1, S2 and S3, respectively; Δt_1 and Δt_2 are time differences $(t_2 - t_1)$ and $(t_3 - t_1)$, and c is the wave speed.

The location of the source is given by

$$x_s = x_1 + d_1 \cos\theta \tag{5}$$

$$y_s = y_1 + d_1 \sin\theta \tag{6}$$

Using a suitable iteration scheme, the values of θ are varied, and the value that minimizes the error between two calculated source locations can be used to identify the location.

TOA method can lead to errors if different modes or different frequency components are recorded at the sensors, resulting in arrival times being calculated on components that have travelled at different velocities (Surgeon and Wevers, 1999). TOA method is also not suitable for dispersive waves in large structures (Ziola and Gorman, 1991). Three-dimensional source localization presents added complexity. New modes can emerge because of reflections from the boundaries or during the encounter of a change in the geometric or material parameters (Kirikera et al., 2007).

Application of Lamb wave modes in source localization is an attractive option in plate-like structures. Recording the arrival time of two modes and using their velocities, a single sensor could be used to find source location. Use of the theory makes it possible to extend AE technology for global or semi-global monitoring, that is monitoring a larger area. But for structures with complex geometric shape, both modal source location technique and TOA method are not effective. Newer source location methods have been proposed, for example a method based on AE energy has been discussed by Nivesrangsan et al. (2007) and a method based on grid of time differences is used by Baxter et al. (2007).

Noise Suppression

AE signals are often masked by noises, arising from traffic or other environment sources as well as from other sources such as rubbing of the parts and loosening of bolts. Difficulty in distinguishing damage growth-related emissions from other background noises has been identified as the most serious obstacle in AE monitoring (Lozev et al., 1997). Different suggestions have been made for noise suppression, these include high-pass frequency filtering (removes low frequency noise), signal threshold filtering (remove low amplitude noise), spatial filtering using guard sensors and analysis of signal characteristics (Daniel et al., 1998). An example of the use of guard sensors can be seen in Figure 9.

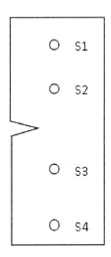

Figure 9. Use of guard sensors

Four sensors, S1, S2, S3 and S4 are placed to record emission from the crack. AE signals from the growing crack are picked up by the sensors S2 and S3 before the other two, as they are closer to the crack. If sensors S1 and S4 pick up signals before sensors S2 and S3, they are likely from sources other than the crack and are rejected.

Analysis of signal characteristics has proved useful in distinguishing among different sources of AE. In steel structures, crack growth and crack face rubbing are found to have much shorter rise times than the noise signals; but crack growth emissions have a much longer duration than crack face rubbing emission (Lozev et al., 1997). Li and Ou (2007) stated that differences in amplitude, duration time, rise time and frequency ranges are effective ways to differentiate among noise, continuous AE waveform and burst AE waveform; and filter and floating threshold and guard sensor can be used to remove ambient noise. Similarly, Sison et al. (1998) have stated that mechanical and fretting noises are easily distinguishable from crack-related AEs because of lower frequency contents and much longer rise times and durations.

Signal Processing

Recording of AE signal waveform is itself not enough; the signals need to be processed for assessment and quantification of damage. Some commonly used signal-processing methods for analyzing AE signals include time series analysis, Fourier transform (FT), Short timed Fourier transform (STFT, also known as Gabor or windowed Fourier transform) and wavelet transform (WT). FT is a commonly used tool to identify the frequency contents of a signal. But the main disadvantage of FT is that information about time of occurrence of frequency components is lost. To obtain both time and frequency information simultaneously,

STFT and WT are both useful. STFT involves multiplying a signal with a short window function and calculating the Fourier transform of the product. The window is then moved to a new position and the calculation is repeated. This gives both time-frequency information of the whole signal. But due to the use of constant window length, resolution is fixed in both time and frequency domain. Compared to fixed length window size of STFT, wavelet analysis uses windowing technique with variable sizes. Long time interval windows are used where more precise low-frequency information is needed, and shorter regions are used where high-frequency information is desired (MathWorks, 2009). Wavelet analysis, thus, breaks a signal into different levels, where each level is associated with a certain band of frequencies in the signals.

WT has been used in areas like AE signal analysis, fracture mode classification of AE signals from composites and detection of a signal in low signal to noise cases (Yoon et al., 2000). Qi et al. (1997) and Qi (2000) used wavelet-based AE analysis to identify signals of different frequency ranges with different failure modes of composites. Wavelet-based filter techniques have proved useful in enhancing signal-noise ratio (SNR) (Chang et al., 2003; Grosse et al., 2004; Grosse et al., 2002). Hilbert-Huang transform is another method that can be used for increasing the SNR. Use of wavelets are useful in other SHM methods, a summary of the use in various SHM applications can be found in Reda Taha et al. (2006).

Source Identification and Severity Assessment

Source identification and damage quantification to assess the severity of sources is another challenge in AE technique. Pattern-recognition techniques are common in various applications of AE monitoring. These involve extracting some features from the signals and using them to classify signals using classifying algorithms such as neural networks. One example is the study by Huguet et al. (2002) where acoustic emission signal parameters and neural network techniques are used to identify and characterize the various damage mechanisms in stressed glass fibre reinforced polymer composite.

Signals from similar sources can be expected to have similar features and frequency content, though the medium of propagation and sensor characteristics can affect the waveforms. The similarity of two signals can be quantified using a mathematical tool called magnitude squared coherence (Grosse and Linzer, 2008). The magnitude squared coherence (MSC) estimate is a function of frequency with values between zero and one that indicates how well two signals correspond to each other at each frequency, with the value of one indicating exact match (MathWorks, 2009). For two signals x and y, MSC is calculated using the power spectral densities (P_{xx} and P_{yy}) and the cross power spectral density of the signals (P_{xy}) as follows:

$$C_{xy}(f) = \frac{\left|P_{xy}(f)\right|^2}{P_{xx}(f) \cdot P_{yy}(f)} \tag{7}$$

Instrumentation for AE Monitoring

Detection, amplification, filtering and analyzing of signals are important issues in the use of AE technique. A typical AE monitoring system consists of sensors, preamplifiers and AE acquisition and analysis system. Sensors used are usually of piezoelectric type. Piezoelectric material convert mechanical waves into electrical signals and vice versa. Operating frequency range is important during sensor selection. The common frequency range for AE testing in steel structures is 100 to 300 KHz (Holford and Lark, 2005). Good coupling of the sensors to the test specimen is necessary for effective transmission of AE signals. Therefore, sensors are attached on the surfaces using magnetic holders, glues or even rubber bands and tapes and a layer of couplant such as vacuum grease and oil is applied between the two surfaces.

AE signals generated are often too small to be detected; hence, preamplifiers are used to amplify the signals before further processing. Typical amplification gain ranges from 40 to 60 dB. Amplified AE signals are fed into AE acquisition system, which along with analysis software, can be used to evaluate the parameters of the signals, process the data and plot the results. A layout of a typical instrumentation can be seen in Figure 10.

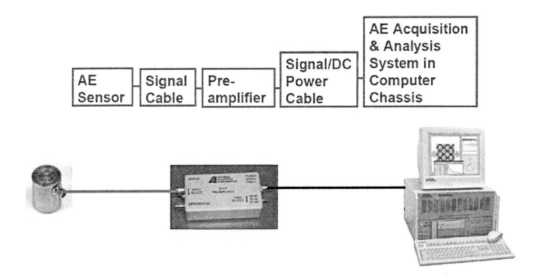

Figure 10. AE instrumentation (PAC, 2007)

SUMMARY OF THE PROJECT

The SHM research group at QUT includes a number of academics and research students. One of the aims of the group is to develop a SHM tool capable of detecting, locating and predicting the severity of damage in bridges by using appropriate sensor technology, dynamic computer simulation techniques and damage assessment models. This project will investigate the use of acoustic emission technique and its integration with other SHM methods and damage assessment models in order to develop an effective SHM tool.

This project will primarily concentrate on steel bridges. Detecting and locating the source of emission is often the first step of any monitoring process. Issues that need to be addressed afterwards include source identification and severity assessment. AE is mainly used as local monitoring technique, so areas identified by visual inspection or other methods or areas susceptible such as welds or jointed connections will be monitored. The main strength of AE technique is in determining if a crack is active or not. Since along with emission from crack initiation or growth other sources of AE are present, strategies need to be developed to address source discrimination.

The project will primarily aim to develop a data analysis model with intelligent use of different signal-processing tools in order to distinguish various sources of AE. Though AE signals recorded by sensors are affected by the sensor characteristics and the propagating path, they still contain information about the source mechanisms. Furthermore, in studies of composites different sources of damage have been found to give signals of different frequency levels. With the use of Short time Fourier transform (STFT) and wavelet analysis, signals are broken into different frequency bands and simultaneous time and frequency representation of signals is possible. This makes it easier to identify high-frequency transient (short-duration) crack signals from lower frequency continuous (long-duration) noise signals. Further, energy contained in different frequency bands of recorded signals can be an effective parameter to discriminate sources and assess their severity. Similarity of recorded waveform signals will also be explored for source discrimination purposes.

EXPERIMENTAL WORK

Background to Experimental Work

The purposes of the experiments were to a) explore accurate source localization by identification of wave modes in plate-like structures, and b) find a way to determine similarity between AE signals from different sources in beam like structures. The AE acquisition system used in the experiments is μ-DiSP PAC (Physical Acoustics Corporation) with four AE channels. AE WIN software supplied by PAC provides easy viewing of waveform and signal parameters and some data post processing. MATLAB (MathWorks, 2009) is used extensively for analyzing data. Data was acquired at a sampling rate of 1 MHz; and, therefore, according to Nyquist theorem, only frequencies up to 500 KHz could be recorded. Sensors used were PAC R15a, resonant at 150 KHz, with prescribed operating range between 50 to 200 kHz.

Primary sources of acoustic emission were pencil lead breaks. Pencil lead break test involves breaking pencil leads on the surface of test specimen and provides a simple way of generating crack-like signals. Filters in the acquisition system were set at 20 KHz and 400 KHz, as pencil lead breaks are generally found to emit signals within this range. A threshold value needs to be set in the acquisition system, and recording is triggered once the output signals reach this value. Threshold set depended on experiments and generally ranged between 45 to 60 dB. Preamplifiers provide an amplification choice of 20 dB, 40 dB or 60 dB. Due to relatively noise free environment in laboratory, an amplification of 20 dB was enough for most experiments. In addition to pencil lead break tests, dropping small steel balls from a set height were also used to act as sources of AE signals in one set of experiments. Beams and plate-like structures are common in bridges and were therefore used as test specimen.

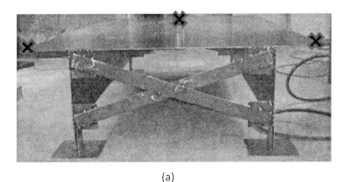

x − Sensor
positions

(a)

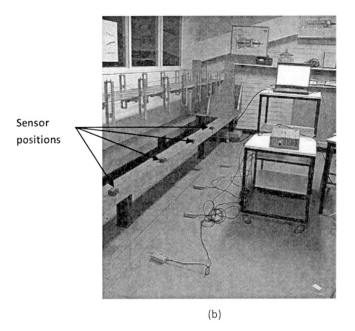

Sensor
positions

(b)

Figure 11. Set up for (a) Source location experiment and (b) Signal similarity experiment

Source Location Experiment

Experiments were carried out in a steel plate, which formed the deck of a slab-on-girder bridge model, and had dimensions of 1.8 m by 1.2 m and thickness of 3 mm, shown in **Figure 11(a)**. The main aim of the experiments was to investigate the influence of wave velocities in source location by TOA method and to identify AE wave modes for accurate determination of source location.

AE signals were generated by breaking 0.5 mm pencil leads at selected locations on the plate. For each position, two pencil lead break tests were carried out. Differences in arrival times of signals at three sensors were calculated manually, and source locations were calculated using longitudinal and transverse wave velocities of waves and then compared with exact locations. Accurate identification of wave modes and exploration of presence of Lamb wave modes was then attempted by frequency and time-frequency analyses of the signals.

Signal Similarity Experiment

A 4 m long steel square channel beam present in a suspension bridge model was used as the test specimen, **Figure 11(b)**. Along with the standard pencil lead breaks, steel ball drop tests were carried out to generate AE signals. The steel balls were 6 mm in diameter and were dropped from a vertical height of 15 cm. Ten sets of each test were carried out. The signals were recorded by a sensor placed at a distance of 1.5 m from the source. They were then analyzed to see if the signals from the two sources could be distinguished by using magnitude squared coherence.

RESULTS AND DISCUSSIONS

Source Location

Locations of the emission sources were calculated using the longitudinal wave velocity ($c_L = \sqrt{E/\rho} = 5188$ m/s) and the transverse velocity ($c_T = \sqrt{E/2\rho(1+\upsilon)} = 3000$ m/s) and results are shown in **Figure 12** with exact positions, calculated positions and the sensor locations as marked. The following values were used for material properties: Young's modulus, $E = 210$ GPa; density, $\rho = 7800$ kg/m³ and Poisson's ratio, $v = 0.3$.

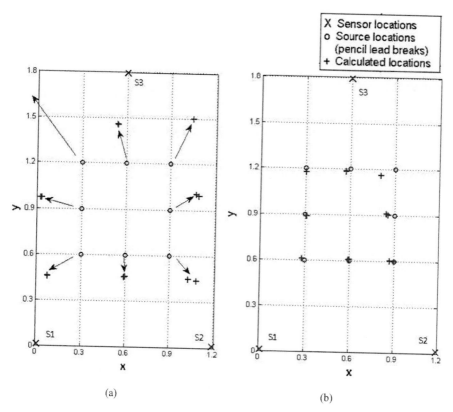

Figure 12. Source location results using (a) longitudinal and (b) transverse wave velocities

It is seen that the use of c_L does not give good match between the exact and calculated values, whereas using c_T gives much better correlation. To identify the wave modes recorded by the sensors, initial part of one sample signal (recorded by Sensor S3 for pencil lead break at position (0.3, 1.2)), was analyzed next. It is shown in **Figure 13**, along with the preset threshold value (dotted line). It is seen that the initial low amplitude signals do not cross the threshold and thus do not trigger a hit (initiation of recording in data acquisition system). It is also observed that the triggering wave component arrives around 90 µs afterwards. Using the velocity of the triggering wave as c = 3000m/s (as this value gave good source location results), the distance between the source and signal (0.67 m) and a time difference of 90 µs; the velocity of the initial wave can be calculated to be around 5000 m/s [= 0.67 m/ (0.67/3000) s - 90.10^{-6} s]. This value is close to the longitudinal velocity of waves in steel. Initial conclusions can be drawn that though longitudinal waves are present, they have attenuated to a level below the threshold; and the waves that record a hit by crossing the threshold are the transverse waves.

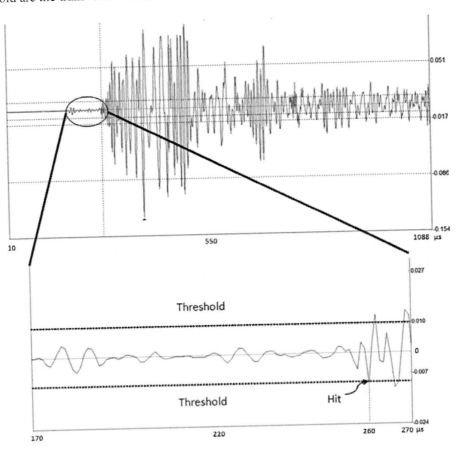

Figure 13. Initial part of sample signal

Lamb waves are common in large plate-like structures, and, hence, further investigation was carried out to check whether the waves seen were the Lamb wave modes, by exploring the frequency contents of the signal. Fourier analysis was carried out separately for two parts of the sample signal in Figure 13: the initial 90 μs portion and the next 230 μs portion. The results are shown in **Figure 14**.

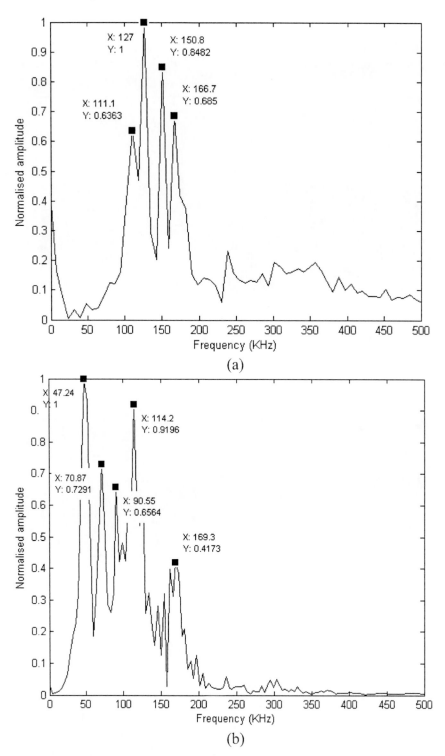

Figure 14. FFT of two portions of signal (a) initial 90 μs and (b) next 230 μs of the signal

The major difference observed between **Figure 14(a)** and **Figure 14(b)** is that frequency peaks exist around 47, 70 and 90 kHz in the latter one, indicating that these lower frequency wave modes arrive late and trigger a hit. To obtain time-frequency information simultaneously, Short time Fourier transform (STFT) analysis was carried out using time-frequency toolbox (Auger et al., 1996). Squared of STFT coefficients are plotted with respect to both time and frequency and plots in linear and logarithmic scales are shown in **Figure 15**.

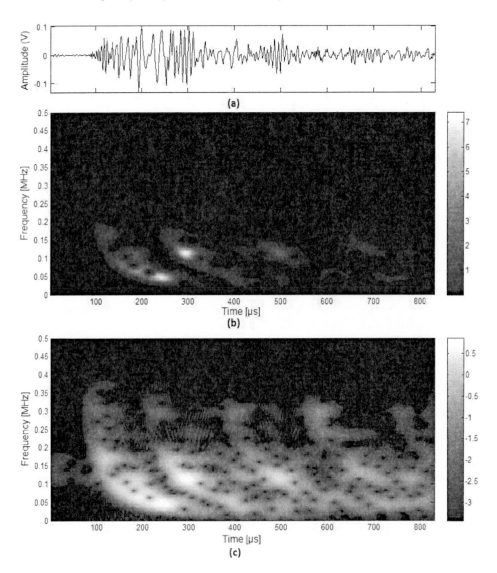

Figure 15. STFT analysis (a) Sample signal (b) in linear scale and (c) in logarithmic scale

Though not clear in linear plot in **Figure 15(a)**, logarithmic plot in **Figure 15(b)** shows that waves with frequencies around 100 to 180 kHz arrive at the beginning. In both plots, it is visible that starting at around 90 μs, waves with a large variation in frequencies arrive. The frequencies gradually decrease from around 350 kHz to 30 kHz, with a peak value at around 45 kHz and 250 μs. Another similar wave pattern with decreasing frequency values emerges near 300μs; these are likely to be reflected waves.

For further insight into Lamb wave phenomenon, the dispersion curve for steel is studied (**Figure 16**). Using the frequency f of 100 kHz to 180 kHz and 3 mm thickness t of the plate ($f{\cdot}t$ = 0.3 to 0.54 MHz·mm), group velocity of around 5000 m/s is seen for S_0 mode. This value matches with the calculations made before for initial fast arriving component. But for the triggering slow arriving wave (of velocity 3000 m/s) to be flexural mode (A_0), frequency of around 333 kHz or more is required ($f{\cdot}t$ = 1 MHz·mm). This high-frequency component, though visible in log STFT plot, is not conspicuous in other plots. Hence, it implies that this mode is weak, and it is unlikely it crossed the threshold first.

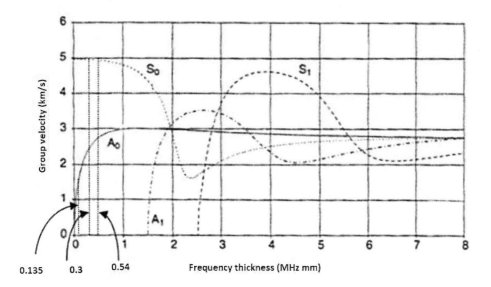

Figure 16. Dispersion curves for steel (Holford and Lark, 2005)

In previous studies, flexural waves have been found generally to be of lower frequencies than extensional waves (Maji et al., 1997; Ziola and Gorman, 1991). Furthermore, pencil lead breaks act as out-of-plane sources and generate signals that are primarily flexural in nature. Therefore, wave mode with high energy and frequency of 45 KHz arriving at around 250 μs, as seen in STFT and wavelet plots, is most likely to be the flexural component A_0. Again, using the velocity of triggering wave as 3000m/s and the distance traveled by signals as 0.67 m, the velocity of the high-energy wave mode A_0 can be calculated to be 1750 m/s. For the wave components of 45 kHz ($f{\cdot}t$ = 0.135 MHz·mm), group velocity of slightly less than 2000 m/s is seen for A_0 mode in dispersion curve in **Figure 16**, and this is close to the calculated value.

Based on the results, it is concluded that the triggering wave mode is transverse wave but the dominant mode is flexural mode. Attempt to use flexural mode for source location is identified as future work.

Signal Similarity

MSC values between signals from all pencil lead break experiments have a mean of 0.78, while the MSC values between pencil lead break and ball drop signals have a mean of 0.38. The values were calculated using MATLAB code mscohere (MathWorks, 2009). These differences are significant, showing the possibility for using magnitude squared coherence for signal classification purposes.

A typical plot of MSC values versus frequencies between two pencil lead break signals is shown in **Figure 17(a)**, and a similar plot between pencil lead break and ball drop signals is shown in **Figure 17(b)**. **Figure 17(a)** indicates close match of frequencies between the signals, especially up to the value of 400 kHz, whereas **Figure 17(b)** indicates less coherence between the two signals.

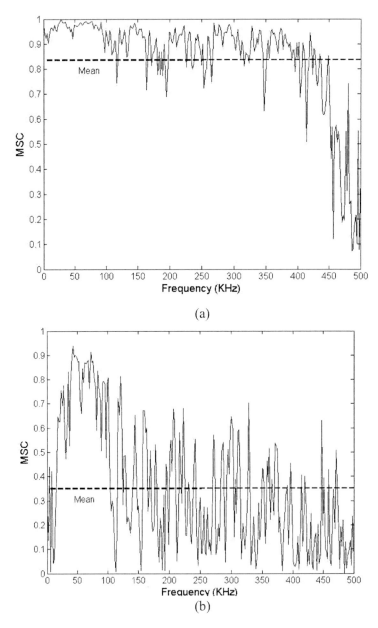

(a)

(b)

Figure 17. MSC values versus frequencies for (a) two pencil lead break sources and (b) a pencil lead break and a ball drop

CONCLUSION

Experiments in this study have explored the use of frequency analysis for better interpretation of recorded AE data. Ability to determine the location of the source is an important advantage of acoustic emission technique. AE propagation in solids is a complex phenomenon, as signals travel in various modes and mode conversions occur. For accurate source location, identification of the modes is necessary, and the results from source location experiments confirm this case. For signal analysis, STFT is found to be more informative than Fourier transform, as both frequencies and times of occurrence of different wave modes are seen in STFT analysis.

Similar sources have been found to emit AE signals with similar waveforms. Magnitude squared coherence (MSC) based on the power spectral frequency analysis of the signals provides a simple way of judging signal similarity, as verified by the experimental results. Signal similarity can be an effective tool for signal classification and thus for source identification and assessment, which are some of the important aspects of AE monitoring method. A crack waveform obtained from a laboratory experiment can act as a template for distinguishing a similar signal obtained in field testing from other noise sources; hence, fatigue testing will be carried out in future. Waveform recorded by a sensor is influenced by the path travelled by the signals and by the sensor characteristics; the influence of these parameters needs further consideration.

The study of acoustic emission technique for monitoring bridge structural integrity is growing continually. AE is a "local" SHM technique and usually involves monitoring of an area with already identified crack to see if crack is actively growing or not. Fracture critical areas such as welded connections or joints can also be monitored to see if they have cracks. Active, that is, propagating, crack will emit AE, while inactive cracks will give no emission. This knowledge is vital as this will help prioritize repair and maintenance work. Another use is in checking whether retrofitting has succeeded in stopping further crack growth. Thus, it is clear that AE technique has several distinct advantages, but it also has several limitations. Proper analysis of recorded AE signals to deduce information about the nature of the source is one challenge. Existence of sources of noises that can mask AE signals from real damage has been identified as a major hindrance for the use of AE technique in monitoring bridges. Signal-processing techniques used in this study can be effective tools of distinguishing and removing noises from real signals.

In AE monitoring applications, a large amount of collected data comes from spurious noise sources, so discrimination of noise sources will mean less data (that is, data only related to crack activity) has to be stored and transmitted. Features from wavelet or STFT analysis (such as energy distribution in different frequency levels) can be effective for pattern recognition tasks. Use of neural network and other tools will be investigated for this purpose in future. Though only laboratory tests have been carried out so far, knowledge from these tests is valuable in interpreting the results from actual field tests.

To summarize, it can be said that effective analysis of recorded data and noise removal still remain big challenges in the use of AE for health monitoring of civil structures such as bridges. However, it is believed the analysis approach by means of time-frequency signal-processing techniques and coherence functions, as applied in this study, can help in better interpretation of experimental data by aiding in source location and in source identification.

AE technique is one of the exciting methods in monitoring structural integrity of bridge structures, and the use of this technique in conjunction with other SHM techniques such as vibration-based technique or other NDT techniques will ensure the bridges perform safely and reliably.

ACKNOWLEDGMENTS

This project is supported by the Australian Postgraduate Award (APA), Australian Research Council (ARC) and Cooperative Research Centre for Integrated Engineering Asset Management (CIEAM) grants.

REFERENCES

Achenbach, J. D. (2009). "Structural health monitoring - What is the prescription?" *Mechanics Research Communications*, 36, 137-142.

Ansari, F. (2007). "Practical implementation of optical fiber sensors in civil structural health monitoring." *Journal of intelligent material systems and structures*, 18, 879-889.

Auger, F., Flandrin, P., Goncalves, P., and Lemoine, O. (1996). "Time-Frequency Toolbox - For use with MATLAB." CNRS (France) and Rice University (USA).

Austroads. (2004). "Guidelines for Bridge management - Structure Information." Austroads Inc, Sydney, Australia.

Baxter, M. G., Pullin, R., Holford, K. M., and Evans, S. L. (2007). "Delta *T* source location for acoustic emission." *Mechanical systems and signal processing*, 21, 1512-1520.

Brownjohn, J. M. W., Moyo, P., Omenzetter, P., and Chakarborty, S. (2005). "Lessons from monitoring the performance of highway bridges." *Structural control and health monitoring*, 12, 227-244.

Carlos, M. (2003). "Acoustic emission: Heeding the warning sounds from materials." American Society for Testing and Materials International (*www.astm.org*).

Chang, P. C., Flatau, A., and Liu, S. C. (2003). "Review paper: Health monitoring of civil infrastructure." *Structural health monitoring*, 2, 257-267.

Chang, P. C., and Liu, S. C. (2003). "Recent research in nondestructive evaluation of civil infrastructures." *Journal of materials in civil engineering*, 298-304.

Daniel, I. M., Luo, J. J., Sifniotopoulos, C. G., and Chun, H. J. (1998). "Acoustic emission monitoring of fatigue damage in metals." *Nondestructive testing evaluation*, 14, 71-87.

Farrar, C. R., Doebling, S. W., and Nix, D. A. (2001). "Vibration-based structural damage identification." *Philosophical Transactions of the Royal Society A*, 359, 131-149.

Finlayson, R. D., Luzio, M. A., Miller, R., and Pollock, A. A. (2003). "Continuous health monitoring of graphite epoxy motorcases (GEM)." *CINDE Journal*, 15-24.

Golaski, L., Gebski, P., and Ono, K. (2002). "Diagnostics of reinforced concrete bridges by acoustic emission." *Journal of acoustic emission*, 20, 83-98.

Gong, Z., Nyborg, E. O., and Oommen, G. (1992). "Acoustic emission monitoring of steel railroad bridges." *Materials Evaluation*, 50(7), 883-887.

Gorman, M. R. (1998). "Some connections between AE testing of large structures and small samples." *Nondestructive testing and evaluation*, 14, 89-104.

Grosse, C. U., Finck, F., Kurz, J. H., and Reinhardt, H. W. (2004). "Improvements of AE technique using wavelet algorithms, coherence functions and automatic data analysis." *Construction and building Materials*, 18, 203-213.

Grosse, C. U., Kruger, M., and Glaser, S. D. (2006). "Wireless acoustic emission sensor networks for structural health monitoring in civil engineering." *9th European conference on non-destructive testing*, Berlin.

Grosse, C. U., and Linzer, L. M. (2008). "Signal-based AE analysis." Acoustic Emission Testing, C. U. Grosse and M. Ohtsu, eds., Springer-Verlag.

Grosse, C. U., Reinhardt, H. W., Motz, M., and Kroplin, B. H. (2002). "Signal conditioning in acoustic emission analysis using wavelets." *NDT.net*, 7(9).

Holford, K. M., Davies, A. W., Pullin, R., and Carter, D. C. (2001). "Damage location in steel bridges by acoustic emission." *Journal of intelligent material systems and structures*, 12, 567-576.

Holford, K. M., and Lark, R. J. (2005). "Acoustic emission testing of bridges." *Inspection and monitoring techniques for bridges and civil structures*. G. Fu, ed., Woodhead Publishing Limited and CRC, 183-215.

Huang, M., Jiang, L., Liaw, P. K., Brooks, C. R., Seeley, R., and Klarstrom, D. L. (1998). "Using acoustic emission in fatigue and fracture materials research." *JOM*, 50(11).

Huguet, S., Godin, N., Gaertner, R., Salmon, L., and Villard, D. (2002). "Use of acoustic emission to identify damage modes in glass fibre reinforced polyester." *Composites Science and Technology*, 62, 1433-1444.

Kirikera, G. R., Shinde, V., Schulz, M. J., Ghoshal, A., Sundaresan, M., and Allemang, R. (2007). "Damage localisation in composite and metallic structures using a structural neural system and simulated acoustic emissions." *Mechanical systems and signal processing*, 21, 280-297.

Li, D., and Ou, J. (2007) "Health diagnosis of arch bridge suspender by acoustic emission technique." *Proc. of SPIE*.

Li, H.-N., Li, D.-S., and Song, G.-B. (2004). "Recent applications of fiber optic sensors to health monitoring in civil engineering." *Engineering structures*, 26, 1647-1657.

Lozev, M. G., Clemena, G. G., Duke, J. C., Jr.,, Sison, M. F., Jr., and Horne, M. R. (1997). "Acoustic emission monitoring of steel bridge members." Virginia transportation research council.

Maji, A. K., Satpathi, D., and Kratochvil, T. (1997). "Acoustic emission source location using lamb wave modes." *Journal of engineering mechanics*, 154-161.

Mancini, S., Tumino, G., and Gaudenzi, P. (2006). "Structural health monitoring for future space vehicles." *Journal of intelligent material systems and structures*, 17, 577-585.

MathWorks. (2009). "Matlab R2009a Help guide."

McKeefry, J., and Shield, C. (1999). "Acoustic emission monitoring of fatigue cracks in steel bridge girders." Minnesota Department of Transportation, Minneapolis.

Melbourne, C., and Tomor, A. K. (2006). "Application of acoustic emission for masonry arch bridges." *Strain - International Journal for strain measurement*, 42, 165-172.

Nivesrangsan, P., Steel, J. A., and Reuben, R. L. (2007). "Source location of acoustic emission in diesel engines." *Mechanical systems and signal processing*, 21, 1103-1114.

PAC. (2007). "PCI-2 based AE system User's manual." Physical Acoustics Corporation, Princeton Junction, NJ.

Qi, G. (2000). "Wavelet-based AE characterization of composite materials." *NDT&E International*, 33, 133-144.

Qi, G., Barhorst, A., Hashemi, J., and Kamala, G. (1997). "Discrete wavelet decomposition of acoustic emission signals from carbon-fiber-reinforced composites." *Composites Science and Technology*, 57, 389-403.

Reda Taha, M. M., Noureldin, A., Lucero, J. L., and Baca, T. J. (2006). "Wavelet transform for structural health monitoring: a compendium of uses and features." *Structural health monitoring*, 5(3), 267-295.

Rens, K. L., Wipf, T. J., and Klaiber, F. W. (1997). "Review of non-destructive evaluation techniques of civil infrastructure." *Journal of performance of constructed facilities*, 11(2), 152-160.

Rizzo, P., and di Scalea, F. L. (2001). "Acoustic emission monitoring of carbon-fiber-reinforced-polymer bridge stay cables in large-scale testing." *Experimental mechanics*, 41(3), 282-290.

Rose, J. R. (1999). *Ultrasonic waves in solid media*, Cambridge University Press.

Shih, H. W., Thambiratnam, D. P., and Chan, T. H. T. (2009). "Vibration-based structural damage detection in flexural members using multi-criteria approach." *Journal of sound and vibration*, 323, 645-661.

Sison, M., Duke, J. C., Jr.,, Lozev, M. G., and Clemena, G. G. (1998). "Analysis of acoustic emissions from a steel bridge hanger." *Research in Nondestructive Analysis*, 10, 123-145.

Sohn, H., Farrar, C. R., Hemez, F. M., Shunk, D. D., Stinemates, D. W., and R., N. B. (2003). "A review of structural health monitoring literature: 1996-2201." Los Alamos National Laboratory.

Surgeon, M., and Wevers, M. (1999). "Modal analysis of acoustic emission signals from CFRP laminates." *NDT&E International*, 32, 311-322.

USDoT. (2006). "Status of the Nation's Highways, Bridges, and Transit: Condition and Performance." U.S. Department of Transportation Federal Highway Administration Federal Transit Administration.

Yoon, D. J., Weiss, W. J., and Shah, S. P. (2000). "Assessing damage in corroded reinforced concrete using acoustic emission." *Journal of engineering mechanics*, 126(3).

Yuyama, S., Yokoyama, K., Niitani, K., Ohtsu, M., and Uomoto, T. (2007). "Detection and evaluation of failures in high-strength tendon of prestressed concrete bridges by acoustic emission." *Construction and building materials*, 21, 491-500.

Ziola, S. M., and Gorman, M. R. (1991). "Source location in thin plates using cross-correlation." *Journal of the Acoustical Society of America*, 90(5), 2551-2556.

In: Structural Health Monitoring in Australia
Editors: Tommy H.T. Chan and D. P. Thambiratnam

ISBN: 978-1-61728-860-9
©2011 Nova Science Publishers, Inc.

Chapter 4

STRUCTURAL HEALTH MONITORING IN UNIVERSITY OF WESTERN AUSTRALIA – FROM RESEARCH TO APPLICATION

Hong Hao[*]

University of Western Australia, Crawley, Australia

ABSTRACT

Structural Health Monitoring (SHM) has been attracting enormous research efforts around the world because it targets monitoring structural conditions to prevent catastrophic failure and to provide quantitative data for engineers and infrastructure owners to design reliable and economical asset management plans. With support from Australian Research Council (ARC), Cooperative Research Center for Integrated Engineering Asset Management (CIEAM) and Main Roads WA, intensive research works have been carried out in the School of Civil and Resource Engineering, the University of Western Australia (UWA), on various aspects of structural condition monitoring. These include sensor development, signal processing techniques, guided wave (GW) propagation methods, vibration-based methods, model-updating methods, and integrated local GW and global vibration-based methods. The performance of these techniques and methods are all affected by unavoidable noises and uncertainties in structural modeling and response measurements. Because uncertainties may significantly affect the reliability of SHM, research efforts have also been spent on modeling and quantifying some uncertainties associated with SHM. This chapter reports some of our research results related to Finite Element (FE) modeling errors, measurement noises and uncertainties associated with operation environments and signal processing techniques and reliabilities of different damage indices for SHM. Methods for modeling these uncertainties in SHM are also presented and discussed. The results presented in this chapter can be used to quantify possible uncertainties for a better SHM.

---

[*] School of Civil and Resource Engineering, the University of Western Australia, 35 Stirling Highway, Crawley WA 6009, Australia (hao@civil.uwa.edu.au)

INTRODUCTION

A large number of civil structures have been in service for many years. Their conditions inevitably vary from the original design specifications owing to deterioration, corrosion, fatigue, accidental and natural loads. To manage the risk of aging structures, it is very important to reliably monitor the structural conditions. In the last three decades, intensive research efforts and significant advancements have been achieved in both hardware and software systems and techniques for SHM. In general, SHM techniques can be classified into two groups, namely local and global approaches. Methods related to stress wave propagation, acoustic emission, X-ray and eddy current, etc., are examples of local approaches because their sensing range is rather small; but they could be very reliable in detecting small damages (Malhotra, 1984; Pessiki and Olson, 1997; Wang, et al., 2009). On the other hand, vibration-based methods are global approaches because in theory, any changes in structural properties will affect structural vibration properties such as vibration frequencies and mode shapes. Vibration-based approaches have been comprehensively summarized in two literature reports published by the Los Alamos National Laboratory (Doebling et al., 1996; Sohn et al. 2004). The primary drawbacks of global vibration-based methods are that they are not very sensitive to small damage. Moreover, to identify structural damage location and severity, both local and global methods also need a benchmark or a high fidelity Finite Element (FE) model, but in practice, such a model is often not available.

In the last few years, with supports from Australian Research Council (ARC), Cooperative Research Center for Integrated Engineering Asset Management (CIEAM) and Main Roads WA, intensive research works have been conducted in the School of Civil and Resource Engineering, the University of Western Australia (UWA), on various techniques of SHM. The works include sensor development, signal processing techniques, Guided Wave (GW) methods, vibration-based methods, and model-updating methods for identifying debonding damage and corrosion damage in reinforced concrete structures; loss of pre-stressed force in pre-stressed concrete beams, shear link damage in steel and concrete composite structures; scouring and fatigue damage of subsea pipelines; and progressive damage of structures under increased loading levels. The methods have also been applied to identifying conditions of a few concrete bridges in Western Australia.

The accuracy of structural condition identification depends on the reliability of structural response measurements, signal processing techniques to extract the required structural parameters, index to define and quantify structural damage, a high-fidelity FE model of the structure, and a model-updating technique to update the FE model to locate and quantify the damage severity. All these steps are associated with noises and uncertainties. For example, measurement noise and FE modeling error are unavoidable. Different signal processing techniques may yield different structural parameters from the same measured data; and different damage index definitions may result in different structural damage identifications. Moreover, changing environmental conditions will affect the structural vibration parameters. For many years, most researchers have concentrated on studying the effects of structural damage on its vibration properties but have paid little attention on the normal variations of structural vibration properties owing to changing environmental conditions and other uncertainties (Peeters et al., 2000). If the effect of these uncertainties on structural vibration properties is larger than or comparable to the effect of structural damage on its vibration properties, the structural damage is difficult to be reliably identified. Even these uncertain effects are relatively insignificant; if they are not properly modeled and analyzed, they might result in false identifications on one hand and on the other hand, render the true damage not identified.

Research efforts have also been spent in UWA to model and quantify possible variations in various structural vibration properties. This chapter presents some of the related research results, which can be used in modeling these uncertainties in SHM analyses for better structural condition identifications. Discussions on the influences of these uncertain variations on SHM results are also given. Some techniques developed to model these uncertain variations in SHM analyses are also briefly introduced in this chapter.

UNCERTAINTIES IN SHM

Uncertainties in SHM may arise from many sources, such as inaccurate response measurements, different signal processing techniques employed to extract structural parameters, different indices used to define and quantify structural conditions and damage, inevitable FE modeling errors and model updating algorithms. This section discusses the influences of some of these uncertainties on SHM and presents some of the research results to quantify some of these uncertainties.

FE Modeling Error

Reliable application of SHM method sometimes depends on an accurate reference model of the structure. In vibration-based approaches, usually FE method is used to construct such a model. Because of discretization and simplification, an FE model of a continuous structure introduces modeling errors. Moreover, an FE model is often created from the design drawings and design material properties. In practice, the structure dimensions, boundary conditions and structural material properties will never be exactly the same as the design values. These will also introduce modeling errors. These errors will affect the predicted structural vibration properties and may lead to misinterpretation of structural conditions.

To demonstrate the discretization error, consider a simply supported beam illustrated in Figure 1. The vibration frequencies of the beam are calculated by exact theoretical approach and FE simulation using commercial code SAP2000. The results are given in Table 1, in which the numbers given in the parenthesis are relative errors with respect to the theoretical predictions. As can be noticed, in this particular example, the numerical simulations always underestimate the vibration frequencies of the beam, and the level of underestimation increases with the mode. Depending on the number of elements used to model this simple beam, the FE modeling errors could be substantial. This discretization error is unavoidable because an FE model represents a continuous structure with discretized elements and some interpolation shape functions.

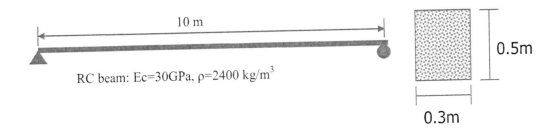

Figure 1. Simply supported concrete beam

Table 1. Theoretical and numerical predicted vibration frequencies of the simply supported beam

Mode	Mode 1 (Hz)	Mode 2 (Hz)	Mode 3 (Hz)
Theoretical	8.02	32.06	72.14
4 elements	7.98 (-0.50%)	30.61 (-4.52%)	65.15 (-9.69%)
5 elements	7.99 (-0.37%)	31.56 (-1.56%)	68.45 (-5.53%)
10 elements	7.99 (-0.37%)	31.68 (-1.19%)	70.13 (-2.79%)

Besides discretization error, configuration error in FE modeling is also very likely owing to simplification in modeling complex boundary conditions, irregular geometry, nonstructural members and secondary structures. As illustrated in Figure 2, a typical bridge girder supported on a rubber bearing with dowel, neither typical support type, roller, pin nor fixed can give this support an exact representation. A soft support with a translational and a rotational spring probably best represents the true condition of the boundary, but the spring constant is not easily determined. Inaccurate modeling of boundary conditions may result in very different structural vibration properties.

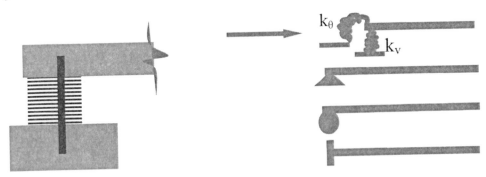

Figure 2. Possible configuration errors in FE modeling by simplifying the boundary conditions

Another possible error in FE modeling arises from the uncertain variations in structural mechanical properties and dimensions. For example, it was found that the coefficient of variation of concrete elastic modulus is about 8% to 10%, and dimension is about 3% (Low and Hao, 2002). Even with very good construction quality control, such variations are inevitable.

To examine the possible variations of concrete structure vibration properties caused by construction quality control, seven continuously supported RC slabs having the same design dimensions and support conditions were constructed in the laboratory. Two slabs were designed according to the linear elastic method and the other five with the moment redistribution method. The seven slabs measured 6400mm x 800mm x 100mm with three meter spans and 200mm overhang on each end. Roller supports were used at both ends of the slab, and the central support was pinned. Grade 32 concrete was chosen, with 20mm aggregate and 80mm slump. N6 reinforcing bar was used for both positive and negative rebar, with a 20mm cover to meet the environmental exposure classification. The reinforcement information is given in Table 2. As can be noted, the reinforcement of the two Linear-Elastic designed slabs is slightly different from the four slabs designed using Moment Redistribution method. However, finite element models indicate the seven slabs have the same vibration properties because such reinforcement difference has little effect on slab vibration properties in a numerical simulation. Six slabs were progressively tested to failure after curing for about 35 days. Another one designed with moment redistribution was casted and left outside the laboratory for three years before destructive testing to monitor the effect of environmental conditions on its vibration properties that will be discussed in the next section.

Table 2. Reinforcement of the six slabs

	Positive	Negative
Linear-Elastic	10 N6-84	13 N6-63
Moment Redistribution	11 N6-75	11 N6-75

Figure 3 shows the reinforcements of the slabs and the load-testing frame. The slabs were designed to take approximately 37.2 kN ultimate load in each span. Vibration test was performed three times on the slabs, i.e., prior to loading, after application of working load and after application of ultimate load. Working load was the loading level that the slabs were designed for. This was 5 kPa, but since the slabs were loaded at third points, the 5 kPa distributed load is equivalent to 23.5 kN per span. At working load, cracks were clearly visible above the central support and in the middle of each span, as shown in Figure 4. Ultimate load was at approximately 40 kN per span and was determined when the testing became unsafe to continue. At ultimate load, the crack widths were very wide (up to 3mm) and the deflection of the slab was clearly visible. Vibration tests at working and ultimate load were performed after the applied loads were removed. For example, a slab was loaded to its working load first, then the load was removed and vibration tests performed. The slab was then reloaded to ultimate load, and load removed and the vibration tests performed again.

A Dytran Model 5802A 5.4kg instrumented impulse hammer was used to excite the slab. The accelerometers used to record acceleration responses are an analogue ADXL 105 ± 5G with a calibration factor of about 850mV/G. Metal plates were glued to the slab using two-part epoxy adhesive to provide a contact surface for the magnetic accelerometers. Metal plates also ensured that the accelerometers were always placed in approximately the same position. Figure 5 shows the slab with metal plates and the test set up. Four thousand and ninety six data points were collected at a sampling rate of 2000Hz giving a Nyquist (cut-off) frequency of 1000Hz. Half a second of data was collected pre-trigger (500 scans) to assess background noise. Channel one of the data logger (attached to the impact hammer) was used as the data collection trigger. Three sets of data were collected for each testing configuration and an average taken.

Figure 3 Slabs and loading frame

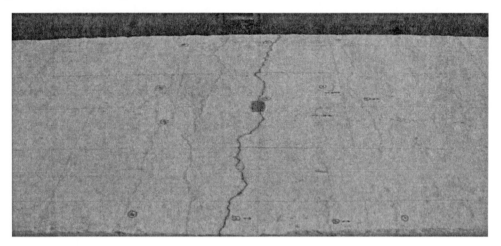

Figure 4. Cracking at working load

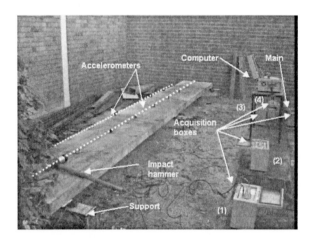

Figure 5. Test set up

The "Rational Fraction Polynomial" (RFP) method (Ewins, 2000) is used to determine the frequencies, the mode shapes, and modal damping of the system. In this study, besides modal damping, the Traditional Logdec Technique (TLT) is also used to estimate the overall damping ratio of the structure from the free vibration decay curves. This technique is applied over successive peaks during the free-vibration phase until the amplitude nears zero, with the damping ratio being averaged to give a fairly accurate estimate of the overall damping. This time domain based technique is commonly used in estimating the damping response of concrete structures. Salzmann et al. (2002) suggest that the TLT method is far superior to any other frequency domain methods due to the low damping levels concrete exhibits. Figure 6 shows the typical measured acceleration time histories from impact testing, and the decay curve of the peak vibrations to determine the equivalent damping ratio.

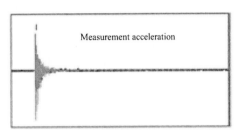 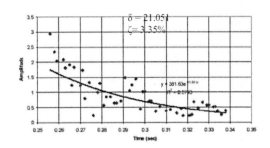

Figure 6. Typical measured vibration time histories and peak value decay curve

Table 3. Measured vibration frequencies of the 6 slabs

Slab	Damage state	Mode 1 (Hz)	Mode 2 (Hz)	Mode 3 (Hz)	Mode 4 (Hz)	Mode 5 (Hz)
1 (Linear-elastic)	Unloaded	18.3	25.4	57.6	67.9	74.7
	Working	16.6	22.5	53.7	62.3	68.6
	Ultimate	13.9	18.1	49.8	58.6	66.9
2 (Linear-elastic)	Unloaded	18.3	26.6	59.6	66.4	73.7
	Working	16.4	21.7	54.9	61.3	68.4
	Ultimate	11.0	15.1	47.6	60.1	67.4
3 (Moment-redistribution)	Unloaded	16.8	23.2	56.6	61.5	70.6
	Working	14.4	20.3	48.8	58.8	65.4
	Ultimate	11.5	14.9	43.2	55.9	59.3
4 (Moment-redistribution)	Unloaded	18.6	25.4	54.9	61.5	67.9
	Working	16.4	21.0	47.9	53.7	60.8
	Ultimate	12.2	15.4	41.7	50.8	56.9
5 (Moment-redistribution)	Unloaded	18.5	24.6	68.2	74.5	88.7
	Working	16.1	21.2	61.3	72.4	87.3
	Ultimate	10.4	15.7	50.2	61.1	71.6
6 (Moment-redistribution)	Unloaded	15.6	24.4	57.1	67.6	82.5
	Working	13.8	17.0	46.9	58.9	73.9
	Ultimate	10.3	15.4	41.3	61.8	88.5

Table 3 lists the vibration frequencies of the six slabs extracted from vibration tests. It should be noted that slab 6 suffered some minor damage with hairline cracks at the center support and in the mid of each span. The damage was caused owing to improper handling during the process of striking off the formwork. From the measured vibration frequencies given in Table 3, these hairline cracks significantly affect the first modal frequency. However, their effect on higher modal frequencies is not prominent. As can be noticed, the vibration frequencies of these supposedly same slabs are different, although they were constructed by the same group of people with the same batch of concrete under the well-controlled laboratory conditions. Table 4 gives the mean values and coefficients of variation of the measured vibration frequencies. As shown, the vibration frequency decreases with the load increment, as expected. The mean frequency reductions for the first five modes are in an order of about 10% to 35% at the working and ultimate loading condition, respectively. The coefficients of variation of the vibration frequencies of the first five unloaded slabs vary from 0.041 to 0.107, and from 0.069 to 0.102, if all the six slabs are included. These observations indicate that the variations of the vibration frequencies of the undamaged slabs are in an order of 4% to 11% for the first five modes. For the unloaded slab, the coefficients of variation increase with the mode, implying the higher the mode, the more significant the variation. These variations are also large enough to affect the reliability of vibration-based structural condition monitoring. In many cases, the change of structural vibration frequency is less than 10%, even when the structure has suffered very prominent damage. The present results indicate that the variations of structural vibration frequencies owing to uncertain structural properties and measurement noise may lead to unreliable structural damage identification if they are not properly accounted for.

Table 4. Mean vibration frequencies and coefficients of variation of the six slabs

			Mode 1	Mode 2	Mode 3	Mode 4	Mode 5
First 5 slabs	Unloaded	Mean	18.1	25.0	59.38	66.4	75.1
		COV	0.041	0.05	0.088	0.081	0.107
	Working	Mean	16.0	21.3	53.3	61.7	70.1
		COV	0.056	0.038	0.101	0.111	0.144
	Ultimate	Mean	11.8	15.8	46.5	57.3	64.4
		COV	0.114	0.082	0.083	0.072	0.095
All 6 slabs	Unloaded	Mean	17.1	24.9	59.1	66.6	76.4
		COV	0.069	0.046	0.081	0.073	0.102
	Working	Mean	15.6	20.6	52.3	61.2	70.7
		COV	0.077	0.093	0.105	0.102	0.130
	Ultimate	Mean	11.6	15.8	45.6	58.1	68.4
		COV	0.117	0.075	0.089	0.071	0.164

The above variations in measured vibration frequencies are not necessarily caused purely because of uncertain structural properties such as dimensions and material properties. Measurement noise, environmental condition change and boundary condition will also affect the structural vibration frequency. The usual approach to overcoming this problem is to update the finite element model in structural damage identification using the measurement data of the undamaged structure. However, vibration data measured from the undamaged structure is often not available. In most cases, the measurement can only be performed on the "current" structure, while the "current" structure may have already suffered certain level of damage.

Table 5. Measured damping ratios of the six slabs

Slab	Damage state	Mode 1 (%)	Mode 2 (%)	Mode 3 (%)	Mode 4 (%)	TLT (%)
1 (Linear-elastic)	Unloaded	7.27	4.12	4.45	2.74	3.01
	Working	8.73	6.48	4.43	3.24	5.93
	Ultimate	9.55	7.09	4.09	3.54	7.39
2 (Linear-elastic)	Unloaded	6.03	4.59	3.63	3.02	3.23
	Working	7.41	7.49	3.51	2.68	4.43
	Ultimate	8.98	9.09	3.58	3.54	8.18
3 (Moment-redistribution)	Unloaded	6.40	5.17	3.79	1.43	2.78
	Working	7.19	5.91	4.80	2.33	3.38
	Ultimate	8.09	7.33	5.12	3.06	8.19
4 (Moment-redistribution)	Unloaded	6.21	4.61	2.62	2.13	2.06
	Working	7.81	7.29	2.47	2.87	2.60
	Ultimate	10.69	9.37	5.14	3.54	6.43
5 (Moment-redistribution)	Unloaded	6.43	5.07	3.32	3.07	3.74
	Working	7.00	6.79	3.74	3.41	4.89
	Ultimate	11.49	8.79	4.60	3.52	8.16
6 (Moment-redistribution)	Unloaded	7.48	5.26	4.17	3.26	3.92
	Working	7.56	7.90	5.39	3.44	4.29
	Ultimate	13.25	9.18	5.54	-	8.53

Damping is another vibration property that is used in structural health monitoring. In general, damping increases with the structural damage. Table 5 lists the damping ratios of the six slabs extracted from the measured data. It should be noted that the fourth modal damping ratio of the slab 6 at the ultimate loading level is not included as this mode did not appear to be excited adequately to reliably extract the damping value. As expected, all the four modal damping ratios and the overall damping ratio estimated from TLT increase with the structural damage. The damping ratios of the six slabs, however, vary from each other significantly. It can be noticed that the extracted damping ratio, especially the first modal damping is rather big. The exact reasons for this abnormally large damping ratio are not known. One possible reason could be the frictions between the slab and the non-idealized supports, especially the center support where the slab simply rested on a rectangular steel hollow section.

Table 6 lists the mean and coefficients of variation of the measured damping ratios of the six slabs. As can be seen, the coefficients of variation of the damping ratios among the six slabs vary from 0.073 to 0.28 for unloaded slabs, 0.09 to 0.238 for the slabs damaged by working load, and 0.062 to 0.138 for the slabs at the ultimate loading level. For the unloaded slabs, the coefficients of variation of the measured damping ratios increase with the mode. However, this is not true for the slabs at ultimate loading level. Again, these variations are very significant and will definitely affect the reliability of structural condition monitoring. For example, for slab 5, the increase of damping ratio from unloaded to working load condition is 8.9% and 11.1%, respectively, for mode 1 and mode 4, but the corresponding coefficient of variations of the damping ratio across the first five unloaded slabs are 7.3% and 28%, indicating the variations caused by uncertain structural properties are comparable (mode 1) or even larger than those caused by structural damage. This will make the structural damage undetectable or false detection (undamaged structure detected as damaged) if these variations are not properly considered.

Table 6. Mean and coefficient of variation of the measured damping ratios of the six slabs

			Mode 1	Mode 2	Mode 3	Mode 4	TLT
First 5 slabs	Unloaded	Mean	6.57	4.71	3.56	2.48	2.96
		COV	0.073	0.093	0.188	0.280	0.208
	Working	Mean	7.63	6.79	3.79	2.91	4.25
		COV	0.090	0.093	0.238	0.149	0.306
	Ultimate	Mean	9.76	8.33	4.51	3.44	7.67
		COV	0.138	0.126	0.151	0.062	0.101
All 6 slabs	Unloaded	Mean	6.64	4.80	3.66	2.61	3.12
		COV	0.089	0.091	0.177	0.268	0.217
	Working	Mean	7.62	6.98	4.06	3.00	4.25
		COV	0.08	0.104	0.222	0.148	0.273
	Ultimate	Mean	10.34	8.48	4.68	-	7.81
		COV	0.181	0.118	0.157	-	0.099

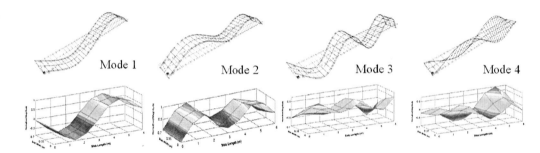

Figure 7. Comparison of FE element simulated and measured vibration mode shapes of slab 1

It should be noted that the measured mode shapes of the six slabs also vary from each other and differ from FE model simulation. Figure 7 shows the comparison of the FE model generated and measured first four mode shapes of slab 1 at undamaged condition. Significant differences can be observed.

The above observations indicate the measured vibration properties of the supposedly same 6 slabs vary owing to uncertain structural properties such as variation in dimensions, material properties and boundary conditions, also owing to possible measurement noises. In particular, for the unloaded slabs, the variations of the measured vibration frequencies are in an order of 4% for the first modal frequency to 11% for the fifth modal frequency, and the variations of the measured modal damping ratios of the unloaded slabs are in an order of 7% to 28%. These variations are significant. If they are not properly modeled, they will affect the reliability of structural condition monitoring.

Influence of Changing Environmental Conditions on Vibration Properties

Changing environmental conditions such as temperature and humidity will cause a change in structural vibration properties. It has been found that the changes of structural vibration properties due to changing environmental conditions could be more significant than those due to rather severe structural damage. For example, it was found that the first three natural frequencies of the Alamosa Canyon Bridge varied about 4.7%, 6.6% and 5.0% during a 24-hour period as temperature of the bridge deck changed by approximately 22°C (Cornwell et al., 1999; Sohn et al., 1999) but a significant artificial cut in I-40 Bridge caused only insignificant frequency changes (Farrar et al., 1997). Peeter and De Roeck (2001) studied the effect of changing environmental conditions on structural vibration properties by monitoring the Z24-Bridge in Switzerland for about a year. They found that the first four vibration frequencies varied 14~18% during the ten months. It was also found that frequencies of all the modes analyzed, except the second mode, decreased with the temperature increase. The second mode frequency, however, increased when the temperature was above 0°C. All the modal vibration frequencies increased significantly when temperature decreased to 0°C and below, probably because of stiffening from the frozen layers. A later progressive damage test of the bridge found that frequencies decreased less than 10% until the final damage scenario (Maeck et al., 2000), indicating changing temperature may have more significant effects on bridge vibration frequencies than damage. Other studies by Askegard and Mossing (1998), Rohrmann and Rucker (1994), Rucker et al. (1995), Khahil et al. (1998), Rohrmann et al. (2000), and Cawley (1997) all concluded that vibration frequencies decrease with temperature increase. Although other vibration properties such as damping and mode shapes are also commonly used as parameters in SHM, the above studies examined only the effects of changing temperature on vibration frequencies. It is known that temperature is the single most important environmental factor affecting the structural damping (Nashif et al., 1985). Changing environmental conditions might also affect the structural vibration mode shapes, but none of the above examined the changing temperature on vibration mode shapes. Moreover, no study has explicitly investigated the changing humidity on structural vibration properties, though its effect is implicitly implied in the papers by Cornwell et al. (1999) and Ciora and Alampalli (2000).

     To study the effect of temperature and humidity on structural vibration properties, a concrete slab as described above was placed outside the laboratory and periodically vibration tested once about every two weeks for three years. During each test, temperature and humidity were recorded. The vibration frequencies, mode shapes and damping ratios were extracted from the testing data and correlated with temperature and humidity to quantify the influences of environmental conditions on structural vibration properties.

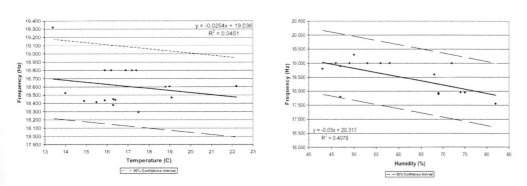

Figure 8. First vibration frequency versus temperature and humidity

Figure 8 shows the typical measured fundamental vibration frequencies against temperature and humidity. As shown, the vibration frequency decreases with the increase of temperature and humidity. Based on the measured data, the following empirical formula is proposed to model the change of vibration frequencies with temperature and humidity.

$$f / f_0 = 1.0 + \beta_t T + \beta_h H \tag{1}$$

where f_0 is a reference frequency of the slab in Hertze. It is the intersection point of the frequency curve with the vertical axis of the graph; β_t and β_h are two coefficients corresponding to temperature and humidity changes and T is temperature in °C. Their values for the slab considered in the study are given in Table 7. It should be noted that these values are only applicable to the temperature range of 10~40°C and humidity of 15~80%, and to the slab under consideration.

Table 7. Coefficients of empirical frequency relation with temperature and humidity

Mode	1	2	3	4
$f_0(Hz)$	19.3154	27.1826	175.2446	98.6824
$\beta_t(°C)$	-0.0023	-0.0023	-0.0013	-0.0050
$\beta_h(°C)$	-0.0003	-0.0004	-0.0002	-0.0011

As can be noticed, the first four modal frequencies of the slab decrease with the temperature and humidity increase; changing temperature has a more significant effect on vibration frequencies than changing humidity. The rate of frequency decrease for mode 1, 2, and 3 are quite similar, whereas that for mode 4 is substantially larger. As shown in Figure 7, mode 1, 2 and 3 are flexural modes and mode 4 is a torsional mode. One possible reason for such differences is that the torsional mode is affected by shear rigidity of the structure, which has different sensitivity to temperature and humidity as bending rigidity does. The results indicate that the bending modal frequencies decrease 0.13~0.23%, while torsional modal frequency decreases 0.5% when temperature increases a unit degree, or respectively decrease 0.02~0.04% and 0.11% for bending and torsional modes when relative humidity increases a unit percent. More detailed discussions, including some analytical predictions of temperature and humidity effects on structural vibration frequencies, are given in reference (Xia et al., 2006).

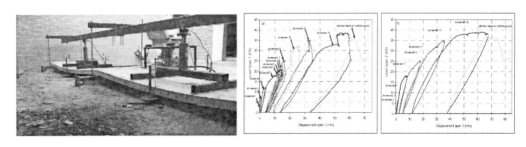

Figure 9. Progressive testing set up and the measured load-displacement relations

After it had been periodically monitored for about three years, the RC slab was progressively tested to failure. Figure 9 shows the test setting up. There are a total of 12 loading stages as shown in Figure 10, in which the induced crack patterns are also shown. During the load testing, two LVDTs were used to measure displacements at the center of two spans. The load-displacement curves of the two spans corresponding to different loading cycles are shown in Figure 9. After each loading application, the load was removed and vibration tests were carried out to extract the slab vibration properties.

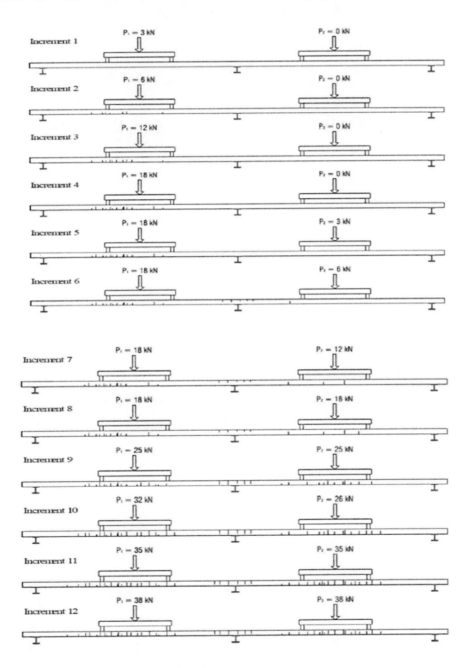

Figure 10. Progressive loading tests and crack patterns

Figure 11 shows the measured reductions in vibration frequencies at different loading stages. As shown, the vibration frequency reduction for most modes, except the fourth torsional mode, is less than 5% up to loading level 5, although obvious cracks on the left span were observed at this loading level. Assume humidity remains unchanged and the rate of frequency change is 0.0023 for the first three bending modes as given in Table 7, a 22 °C temperature change will cause a 5% change in vibration frequency. These results demonstrate that if the influences of varying environmental conditions on structural vibration properties are not properly modelled, using frequency change in SHM may not lead to accurate damage identifications.

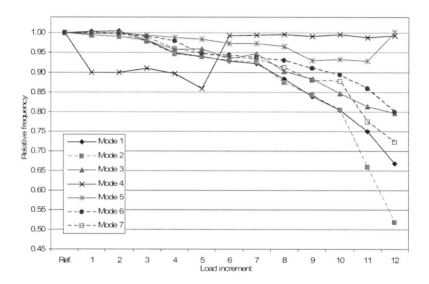

Figure 11. Progressive frequency change corresponding to different loading levels

Similar to vibration frequencies, the influences of changing environmental conditions and structural damage on damping ratios are also studied. Figure 12 shows the typical damping ratio change with respect to temperature and humidity.

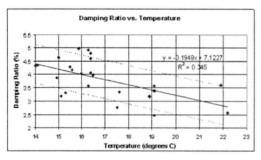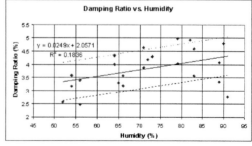

Figure 12. Damping ratio versus temperature and humidity

To quantify the temperature and humidity effects on damping ratio, a linear regression model is derived as

$$\xi = \alpha_0 + \alpha_t t + \alpha_h h \tag{2}$$

where ξ is the percentage damping ratio and α_0, α_t and α_h are regression coefficients. The least-squares fitted α_t and α_h values are given in Table 8. The R^2 statistics are 0.0095, 0.0549, 0.1977 and 0.0653, indicating very poor fittings because the damping ratio varies rather randomly with temperature and humidity. Unlike vibration frequencies, the coefficients in Table 8 indicate that increasing temperature and humidity results in a decrease in the first modal damping, but an increase in the second to fourth modal damping. These observations indicate that it is difficult to quantify the variations of structural damping ratios with temperature and humidity.

Table 8. Coefficients of empirical damping ratio relation with temperature and humidity

Mode	1	2	3	4
α_t(°C)	-0.0034	0.0113	0.0837	0.0180
α_h(°C)	-0.0009	0.0018	0.0144	0.0049

Damping ratios of the slab corresponding to the 12 loading stages were also extracted during the progressive loading tests. Figure 13 shows the typical modal damping ratios corresponding to the loading levels. As shown, the damping ratio generally increases with the structural damage level. However, the increment is prominent only in some modal damping ratios such as that of the first and second modes, but it is insignificant for other modal damping. These observations indicate that as compared to the vibration frequency, damping ratio is not a good parameter to indicate structural damage. It is also more difficult to quantify the changing environmental effects on damping ratios. Therefore, changing damping ratio should be used more cautiously for SHM.

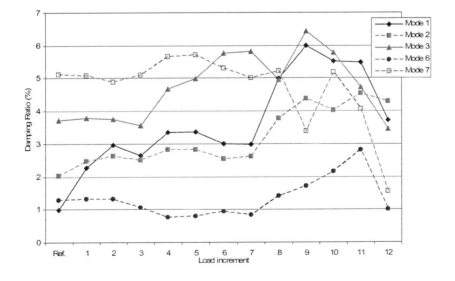

Figure 13. Damping ratio versus progressive loading levels

Effects of changing temperature and humidity on vibration mode shapes were also examined. It was found that changing temperature and humidity has insignificant influence on *MAC and COMAC* values of the studied RC slab. The *MAC* values for Mode 1, 2 and 3 are between 0.988 and 1, that of Mode 4 varies between 0.75 and 1.0, only because the torsional mode was not reliably measured as discussed above. This is because the changing temperature and humidity are rather uniform across the entire the slab. This observation might not be valid for other structures, e.g., a thick structure with temperature and humidity gradients. More discussions regarding the effects of changing environmental conditions on *MAC* and *COMAC* values can be found in (Xia et al. 2006).

Uncertainties Associated with Ambient Vibration Tests for Extracting Structural Properties

A lot of research effort has been spent on developing algorithms to reliably extract structural parameters from ambient vibration tests. Some methods, such as the Stochastic Subspace Identification method (SSI) (Van and De Moor, 1996) and Frequency Domain Decomposition (FDD) method (Brincker et al., 2000), have been developed and successfully used in vibration tests. These methods were developed for linear system identification and are based on time domain measurements and covariance or coherency matrices of the measured dynamic responses. The primary limitation is that they assume the unknown ambient vibration force is a white noise process. In reality, ambient vibration source such as wind, traffic and sea wave may have dominant frequency bands, and the noise in the measured responses is not necessarily white. These may lead to some errors in the extracted structural vibration parameters. Another possible source of error is that they were developed with the linear response assumption, whereas the structure system is often nonlinear after damage occurs.

A few researchers have investigated the reliability of ambient vibration tests on extracting structural vibration parameters. Doebling et al. (1997) compared the impact hammer and ambient vibration tests on the Alamosa Canyon Bridge and found that the impact hammer test identified more modes than the ambient test. The common mode shapes identified from both methods were well correlated. The percentage difference in modal frequencies ranged from 0.03% to 3.19%, whereas the percentage difference in damping ratios ranged from 18.21% to 109.33%, indicating the damping ratio was affected by the type of excitation more significantly than the modal frequency. Peeters et al. (2000) compared the results from shaker, drop weight and ambient excitation of the Z24 Bridge. Ten modes were identified using shaker excitation, but the ninth mode could not be identified from the drop weight excitation and neither the seventh nor tenth mode could be identified from the ambient excitation. Comparing the drop weight and ambient excitation with shaker test, the percentage difference between the frequencies was negligible, but the maximum percentage difference in the ambient damping ratio was 53.06%, larger than 22.5% from the drop weight test. There was good correlation between the first five mode shapes for all sources of excitation, with a Modal Assurance Criteria (MAC) value above 0.84. Recently, Xia et al. (2007) conducted both ambient and multiple reference impact testing on a newly constructed bridge in Perth Western Australia before its opening. Contrary to the above two papers, the ambient tests identified ten modes between 6 and 27 Hz, whereas the impact test failed to identify mode 9. Ambient test in general gave higher frequencies and larger damping ratios than the impact test. The identified frequencies from both tests are comparable, with a difference ranging from 0.35% to 2.70%, but the identified damping ratios differ significantly, varying from 3.45% to 156.10%.

To further study the applicability of ambient vibration tests and to quantify the differences between the vibration parameters obtained from different tests, three scaled models representing a three-storey building, an offshore platform and a pipeline were tested by forced and ambient excitations. The excitation sources include impact force from an instrumented impact hammer, traffic induced-ground motion, and wind pressure generated in a wind tunnel and water wave in a flume or a wave tank. The extracted vibration properties from the forced vibration tests are used as benchmark and compared to those from the ambient tests. Discussions on the differences between the results from different tests are made. Figure 14 shows the test model in different environments.

A modified DIAMOND program originated from Los Almos Laboratory was used to extract vibration properties from the forced vibration test. Two in-house programs based on SSI and FDD algorithms are used to processing the ambient vibration test data to extract structural vibration properties. The differences between the identified vibration properties of the three-storey building model are presented and discussed. Table 9 gives the percentage difference of the modal frequencies from the ambient tests with respect to the impact hammer tests. It can be noticed that wind tunnel tests identified all the seven modes as the forced vibration tests, but the largest frequency difference is more than 8%. Ambient vibration tests based on traffic-induced ground vibrations only identified four out of the seven modes, and the frequency difference varies from 0.15% to 3.72%. Unlike the observations made by Peeters et al. (2000) that the differences between the identified frequencies from forced and ambient vibration tests increases with the mode number, the current test results do not show a clear trend. For example, comparing the frequencies extracted from impact hammer and wind tunnel tests, the largest difference occurs at the second modal frequency and the smallest difference at the fourth modal frequency.

Figure 14. Scaled model tests under different excitations: a) Platform model in wind tunnel; b) Platform model in flume; c) Building model exposed to traffic; d) Impact tests on building model; e) Pipeline model in wind tunnel

Table 9 Percentage difference between modal frequencies of the building model from different tests

Mode	1	2	3	4	5	6	7
Traffic	2.54%	-	-	-0.15%	1.12%	-	-3.72%
Wind Tunnel	1.11%	-8.32%	-1.02%	-0.41%	0.96%	-5.67%	-0.50%

Tables 10 and 11 give the percentage errors of the identified modal damping ratios and MAC values. As can be noticed, the impact hammer tests give consistently higher damping identification as compared to the ambient vibration tests. The identified damping ratios from ambient vibration tests are only about half of those from forced vibration tests. The first six mode shapes identified from the three tests correlate with each other well. However, mode shape 7 identified from the traffic-induced ground motion excitations is very different from that from the forced vibration test, with a MAC value of 0.19, indicating the two identified modes are not correlated at all.

Table 10. Percentage difference between damping ratios of the building model from different tests

Mode	1	2	3	4	5	6	7
Traffic	-98.41%	-	-	-99.69%	-99.87%	-	-99.95%
Wind tunnel	-98.41%	-98.36%	-99.61%	-99.54%	-99.87%	-99.98%	-99.97%

Table 11. Percentage difference between MAC values of the building model from different tests

Mode	1	2	3	4	5	6	7
Traffic	0.97	-	-	0.88	0.95	-	0.19
Wind Tunnel	0.97	0.92	0.86	0.83	0.98	0.93	0.64

Tables 12 and 13 give the percentage differences of the extracted vibration frequencies and damping ratios of the pipeline model obtained by ambient vibration tests in wind tunnel and wave tank against those obtained by impact hammer test. As shown, the vibration frequencies extracted from the ambient vibration tests in wind tunnel and wave tank are in general smaller (except mode 5 from wind tunnel test) than those obtained by impact hammer tests, especially those in wave tank tests where the pipeline model is submerged into water. This observation indicates the fluid-structure interaction reduces the structural vibration frequencies because of the added masses. The largest difference is 2.59% when the pipeline is tested in wind tunnel and more than 8% when tested in wave tank.

Table 12 Percentage difference between modal frequencies of the pipeline model from different tests

Mode	1	2	3	4	5	6
Wind Tunnel	-2.59%	-0.36%	-0.13%	-0.80%	1.80%	-1.52%
Wave Tank	-8.63%	-7.14%	-5.57%	-6.40%	-4.69%	-0.67%

The extracted damping ratios from the ambient tests differ significantly from those by impact hammer tests. As compared to the impact hammer test, damping ratios extracted from ambient tests may differ by more than 200%.

These results demonstrate that structural vibration properties extracted under different operation environments are not exactly the same. The differences in extracted damping ratios in different operation environments are significant, which makes the use of damping ratio in

SHM rather difficult unless a reliable reference model is available. The extracted vibration frequencies and mode shapes also differ, but at a substantially smaller scale. The largest difference in vibration frequencies could be more than 8%, and the smallest MAC value is 0.83. These differences, if not properly modelled, will lead to erroneous structural condition identification.

Table 13 Percentage difference between damping ratios of the pipeline model from different tests

Mode	1	2	3	4	5	6
Wind Tunnel	-27.08%	-32.12%	-11.76%	1.79%	156.76%	85.19%
Wave Tank	-58.33%	16.79%	-5.88%	-10.18%	129.73%	233.33%

Another possible source of errors arises from the data processing techniques. Some preliminary studies have been carried out in UWA, and they found that different data-processing techniques may result in different structural vibration properties. For example, the same ambient vibration test data of the offshore platform model have been processed by using FDD, SSI, and Blind Source Separation (BSS) method. Different vibration frequencies and damping ratios were extracted. In particular, FDD method fails to identify a number of modes when the model was tested in the flume with current force. This is because the FDD method assumes the unknown excitation force as a white noise, but the water current corresponds to a strong harmonic force. Nonetheless, SSI method performs better than FDD, although it also assumes the excitation force is a white noise, but it works in the time domain instead of the frequency domain as FDD method. BSS method gives the best structural vibration modal properties extraction among the three methods examined so far. More detailed results regarding the possible variations in extracted structural vibration properties by using different techniques will be reported in the future.

SENSITIVITY AND RELIABILITY OF DIFFERENT INDICES FOR SHM

Numerous structural vibration parameters have been used in structural condition monitoring and damage detection. Each of these parameters has various advantages and disadvantages. Many researchers compared the sensitivity and reliability of using these parameters for structural condition monitoring. Most of these comparisons are based on either the results of numerical modeling or the testing of a simple experimental specimen with idealized boundary conditions, abundant accelerometers, and simplified damage scenarios. Zhao and DeWolf (1999) conducted a sensitivity study of the ability of natural frequencies, mode shapes and modal flexibilities in detecting structural damage. They used numerically simulated data from a spring-mass system in the analyses and found that modal flexibilities are most sensitive as compared to modal frequencies and mode shapes. Using the testing data from I-40 bridge, Farrar and Jauregui (1996) analyzed the sensitivity of strain energy, modal curvature, flexibility, and uniform flexibility shape curvature for damage identification and found that all methods were able to accurately locate the most severe damage scenarios considered, but the methods were inconsistent and did not clearly identify the damage location for the less severe damage cases. The strain energy method appeared to be the most successful. Salawu and Williams (1995) carried out the dynamic testing of a multispan reinforced concrete highway bridge before and after localized strengthening works. They found that the strengthening resulted in a slight reduction in the natural frequencies and no definite trend for the change in damping ratios. MAC and COMAC values were able to provide information about the location of the repairs. Ndambi et al. (2002) carry out dynamic testing of reinforced concrete beams at various stages of damage by progressively increasing static loads to the specimens. They found that MAC values were less sensitive to damage than natural frequencies. COMAC values were able to detect and locate the damage; however, it was difficult to estimate the damage severity or detect multiple damages. The flexibility matrix was able to successfully monitor the damage severity but was not able to locate damage. The strain energy appears to be the most reliable method for locating the damage. Bayissa and Haritos (2005a) used the mean square value (*MSV*) of the response to measure the total response energy over the entire frequency range for damage detection. They considered a numerical model of a simply supported beam with uniform (global) damage throughout the specimen and localized damage at the mid-span and found that as compared to frequencies, mode shapes, modal flexibility, uniform load surface, modal curvature, flexibility curvature and uniform load curvature, the *MSV* of the response was able to detect damage with sensitivity equal to, or greater than, all of the methods considered. Bayissa and Haritos (2005b) presented additional parameters related to the higher order spectral moments (up to fourth order) for use in damage detection. The study found that the zero-order moment of the power spectral density (*MSV*) was the most accurate to detect and locate damage.

As reviewed above, different studies seem not to give same conclusions regarding the sensitivity and reliability of the various parameters for SHM. To further investigate the sensitivity and reliability of various vibration parameters for SHM, the test data of the two-span RC slab described above are analyzed. Parameters considered include vibration frequency, damping ratio, MAC and COMAC values, modal curvature, modal flexibility, FRF and spectral density function.

As shown in Figure 11, the frequency displays a general decrease due to damage, except the fourth torsional mode. If the torsional mode is omitted from the discussions, there is little variation in the frequencies as a result of the first two load increments, with a noticeable drop occurring after loading level 3, due to the creation of damage in the first span. Additionally, only minor changes are observed between loading level 4 and 6, as the loading of the second span only produces minor cracks in the span. Loading increments eight to ten results in a rapid reduction in the frequencies due to the severe damage produced. The second mode exhibits the largest reduction, especially in the final two load increments. A natural frequency change of 5% has been suggested as a necessary condition to detect the presence of damage with confidence in civil engineering structures (Salawu, 1997). The results shown in Figure 11 indicate that a frequency decrease of 5% is observed in most modes after loading level 4, corresponding to major cracking in the first span. A further 5% reduction of the natural frequencies is observed upon loading the second span with an equivalent load (level 8). These observations imply that the frequency is not very sensitive to small damage because prominent cracks were already observed after loading level 4. However, it is a valuable parameter in SHM because frequencies can be determined easily and reliably. But they should be used with caution to avoid erroneous structural condition identification because changing environmental conditions may also result in similar level of frequency change.

The modal damping ratios for the seven modes shown in Figure 13 indicate that the damping ratios for the bending modes exhibit an increasing trend with damage; however, the trend is not monotonic, with all modes exhibiting a decrease in the damping ratios near ultimate failure of the specimen due to the debonding between the concrete and the steel in the vicinity of the cracks. The first mode of vibration exhibits a relatively large increase in damping ratio as a result of the first two loading increments. Visually, no cracks were observed for the first loading increment, and only a small number of fine cracks were observed for the second increment. This indicates the sensitivity of damping ratio to minor damage in the structure. Similar observations were made by Modena et al. (1999). The damping ratios for the second mode also increased over these increments; however, the observed increase is much smaller. The changes of the other modal damping are not monotonic. Despite the fact that damping is sensitive to small damage, the relationship between damping ratios and damage is complex, and damping ratios are difficult to be accurately determined experimentally. These limit the application of damping as a structural condition monitoring parameter. Nonetheless, damping may provide useful information due to its ability to detect nonlinear dissipative effects as a result of damage (Montalvão et al. 2006; Sohn et al. 2003).

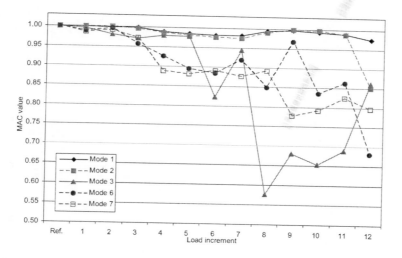

Figure 15. MAC value of the RC slab at different loading stage

The MAC values between damaged and undamaged mode pairs are calculated and shown in Figure 15. As shown, the MAC values of the first mode reduce with the formation of cracks within the first span, successfully indicating the presence of damage. However, the creation of symmetric damage in the second span (loading level 8 onwards) results in the MAC values increasing. This is because the symmetric damage produces similar mode shapes to the undamaged case, as also observed by others (Ndambi et al. 2002; Xia et al. 2006). The higher frequency modes (modes 3, 6 and 7) appear to be more successful at detecting the damage. However, the variation of MAC values with increasing damage is not consistent, as some modes exhibit an increase in the MAC values between certain loading increments, while the MAC values for other modes decrease.

Figure 16 shows the COMAC values for each degree-of-freedom calculated using the undamaged and damaged mode pairs. They show that as the first span is loaded (levels 1 to 4), the COMAC values gradually decrease, with the values in the second span remaining close to unity, successfully locating damage within the first span. However, the loading of the second span (levels 5 to 8) does not display significant changes in the COMAC values in the second span, while the values in the first span increase, indicating failure in detecting damage in the second span. Again, this is explained by the low sensitivity of MAC and COMAC in detecting uniform damage. Ndambi et al. (2002) also experienced difficulty in detecting a second damage region in a reinforced concrete beam if it produced a symmetric distribution of damage.

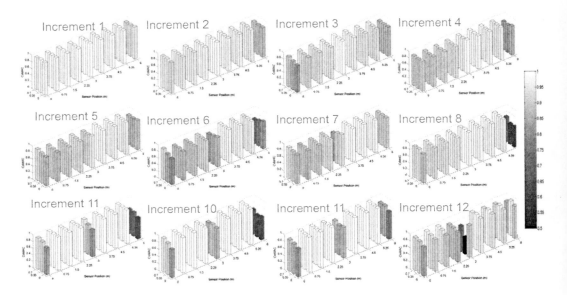

Figure 16. COMAC value at different loading stage

Figure 17 shows the absolute difference between the curvatures of the damaged and undamaged mode shapes for the second mode. Positions 1 and 9 correspond to the outer support of the first and second spans, respectively, and position 5 to the central support. Damage is expected to result in an increase in the curvature difference at the location of the damage, where the magnitude of the curvature difference provides an indication of damage severity (Pandey et al. 1991). As shown, the first four loading increments exhibit large curvature differences in both spans, with a substantial reduction observed for the subsequent four cases (increments 5 to 8). Increments 5 to 8 are remarkably similar to one another, all exhibiting maxima in the centre of the second span, even though damage is progressively increasing in this location. The final four increments all display a further reduction in the curvature difference.

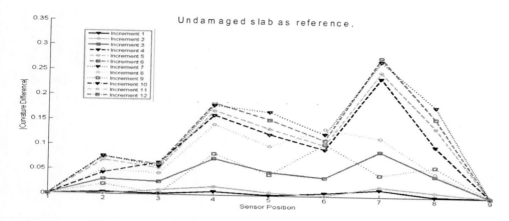

Figure 17.Mode shape curvature at different loading level

Based on the above results, it is unlikely that the mode shape curvature can be used to successfully locate damage (or even detect the presence of damage) in a practical structural condition monitoring application. The insensitivity of this method may be due to the need to calculate the second derivative of the mode shape using a finite difference approximation, which is known to enhance the experimental variability of mode shapes

The modal flexibility matrix (F) was calculated using only the first two modes because the modal contribution to the flexibility matrix decreases with increasing frequency. The results indicate that the flexibility change successfully detects the presence and severity of damage within both spans of the specimen. However, it provides no indication of the presence of damage at the central support. This is due to the small values in the flexibility matrix at the support as a result of it being a nodal point in the mode shapes. To overcome this, a normalized flexibility change parameter ($\hat{\delta}_i$) is proposed as

$$\hat{\delta}_i = \frac{\delta_i}{\max_i |F_{undamaged}|} = \frac{\max_i |F_{damaged} - F_{undamaged}|}{\max_i |F_{undamaged}|} \qquad (3)$$

Figure 18 shows the variation of the normalized flexibility change parameter with damage level. Load increments 1 and 2 result in minor changes in the normalized flexibility change parameter. The formation of cracks within the first span (increment 3) is accompanied by an increase in the normalized flexibility parameter at this location. A further increase is observed as the load within the first span is increased (Increment 4), indicating that the parameter successfully locates damage when the first span is damaged. Increments 5 to 7 correspond to the initial loading of the second span and result in slight increases in the normalized flexibility parameter within first span (as the damage level increases slightly due to the reloading of the damaged specimen), while the values in the second span remain unchanged. The normalized flexibility parameter increases substantially within the second span after load increment 8, successfully detecting the deterioration of the span. Load increment 9 results in further increases in the normalized flexibility parameter in both spans. Additional loading of both spans (increments 10 to 12) results in an increase in the normalized flexibility parameter at all locations (including at the central support) indicating the formation of uniform damage throughout the specimen. The results presented demonstrate that the normalized flexibility parameter successfully identified the existence and location of damage within the specimen.

It is interesting to note the large normalized flexibility change parameter at the outer supports. These are caused by support movements as shown in Figure 18. During the testing, significant support movement was observed. Normalization is also believed to introduce errors because the flexibility at the support is small.

The mean square value (*MSV*) of the response was calculated using the response auto-power spectral densities (Bayissa and Haritos, 2005a). The damage index (*DI*) calculated from *MSV* for each sensor location exhibits no correlation between the damage state and the damage index. This is due to the varying input force provided by the impact hammer for each dynamic test. In order to account for the varying input force used in the different tests, an alternative approach based on the root mean square (*RMS*) value of the frequency response function was considered (Liberatore and Carman, 2004). The frequency response function was squared to account for the complex values. The numerical results indicated that if the entire frequency band was considered in calculating *RMS*, there appears to be no relationship between the *DI* values and the damage severities or locations. This is because the sensitivity of the *RMS* is reduced due to the inclusion of the entire frequency range in the calculation. To overcome this, the *RMS* values of the FRFs in the vicinity of the first mode of vibration were calculated. The results display a clear reduction in the *RMS* values as the damage level increases. The *RMS* values presented indicate that the area under the frequency response function for a carefully chosen frequency bandwidth can detect the presence of damage within a structure, but not the damage location. Similar observation was also made by Liberatore and Carman (2004).

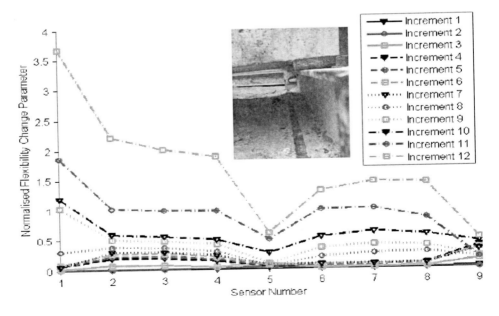

Figure 18. Normalized flexibility change parameter at different loading level

The above results indicate that among all the parameters analyzed, the normalized flexibility change parameter is most successful in detecting the damage in the tested RC slab. However, it should be noted that this conclusion is drawn based only on the tested data of the RC slab under consideration. It may not be true for other type of structures. More studies are definitely needed.

MODELING UNCERTAINTIES IN SHM

The above results demonstrated the uncertain variations in both FE modelling and measured structural vibration properties. If these uncertainties are not properly addressed, they might result in false structural damage identification, i.e., undamaged structure detected as damaged or true structural damage undetected. Significant research effort has been spent on modeling these uncertainties in SHM. Ricles and Kosmatka (1992) assumed the measurement errors and structural parameters are zero mean normally distributed random variable in sensitivity-based model updating calculations. A random error vector is added to the sensitivity equation, and the updating calculation is performed by minimizing the difference in variances of structural parameters. Using perturbation method, a statistical model updating method was developed by considering the measurement error only (Liu, 1995), or modeling frequency errors only (Papadopoulos and Garcia, 1998). Xia and Hao (2003) consider both the FE modelling error and measurement noise in statistical model updating calculations to estimate the probability of damage existence. Farrar and Doebling (1998) applied both Monte Carlo simulation and Bootstrap technique to simulate random data sets based on the statistics of the measured data and use the simulated data in model updating calculations to identify structural properties.

Recently, research work has been carried out in UWA to identify structural conditions using artificial neural network (ANN) with consideration of modelling error and measurement noise (Bakhary et al., 2007). It was found that considering the uncertainties yields better structural condition monitoring results. The method is briefly introduced in the following.

Let the model and measured vibration frequencies and mode shapes equal to the true value plus a random variation as

$$\lambda_i = \lambda_i^0 + \lambda_i^0 X_{\lambda i} = \lambda_i^0 (1 + X_{\lambda i}) \tag{4}$$

$$\hat{\lambda}_i = \hat{\lambda}_i^0 + \hat{\lambda}_i^0 X_{\lambda i} = \hat{\lambda}_i^0 (1 + X_{\lambda i}) \tag{5}$$

$$\phi_i = \phi_i^0 + \phi_i^0 X_{\phi i} = \phi_i^0 (1 + X_{\phi i}) \tag{6}$$

$$\hat{\phi}_i = \hat{\phi}_i^0 + \hat{\phi}_i^0 X_{\phi i} = \hat{\phi}_i^0 (1 + X_{\phi i}) \tag{7}$$

where λ_i, ϕ_i and $\hat{\lambda}_i, \hat{\phi}_i$ are the ith frequencies and mode shapes for training and testing the ANN model, respectively; superscript '0' represents the corresponding mean value, and $X_{\lambda i}$ and $X_{\phi i}$ are the zero mean normally distributed random error in frequencies and mode shapes, which are assumed to be the same for both training and testing data. To develop an ANN model to identify damage in a given structure, FE model of the structure is established first to simulate a large set of structural vibration data corresponding to the structure with different damage scenarios. The simulated vibration frequencies and mode shapes are then smeared with random variations and used to train the ANN model. The trained ANN model relates structural conditions, e.g., stiffness of each element with the structural vibration frequencies and mode shapes. With the well-trained ANN model, the measured vibration frequencies and mode shapes smeared with random variations can be used to identify structural conditions.

Assuming the random variations in Equations (4)-(7) have zero mean and normal distributions, the above procedures can be easily performed with Monte Carlo simulations. The Monte Carlo simulation will give mean, standard deviation and distribution of stiffness parameter of each element. The shortcoming of the Monte Carlo simulation is that it requires a large number of iterations involving training and testing process. Therefore, it is often times consuming and sometime computationally prohibitive. To cope with this problem, the Rosenblueth's point estimate method (Rosenblueth, 1975) is used to approximate the probability moments of damaged and undamaged states of structural parameters. The reliability of the Rosenblueth's point estimate method is verified by Monte Carlo simulation. The Monte Carlo simulation results are also used to derive the statistical distribution, and the observed distribution type is verified by using Kolmorogov-Smirnov goodness of fit test (K-S test). The Rosenblueth's point estimate method is briefly introduced below.

Assuming that there are two random variables X_1 and X_2 in a function $F = fn(X_1, X_2)$, and the prescribed standard deviations for X_1 and X_2 are σ_1 and σ_2 respectively. The mean and standard deviation of F can be obtained through the following steps:

1. Calculate the upper and lower limit of F using the combinations of mean plus one standard deviation and mean minus one standard deviation of each random variable.

$$F_{++} = fn(X_1 + \sigma_1, X_2 + \sigma_2) \tag{8}$$

$$F_{+-} = fn(X_1 + \sigma_1, X_2 - \sigma_2) \tag{9}$$

$$F_{-+} = fn(X_1 - \sigma_1, X_2 + \sigma_2) \tag{10}$$

$$F_{--} = fn(X_1 - \sigma_1, X_2 - \sigma_2) \tag{11}$$

2. Calculate the expectation of F by

$$\mu_F = E(F) = \frac{1}{h}(F_{++} + F_{+-} + F_{-+} + F_{--})$$ (12)

where h is the number of upper and lower limit of F.

3. Calculate the standard deviation, σ_F.

$$\sigma_F = \left[E(F^2) - (E(F))^2\right]^{1/2}$$ (13)

where $E(F^2)$ is calculated using Equation (12) with F^2 terms substituted for the F terms.

In statistical ANN, the upper and lower limit can be obtained by using mean plus one standard deviation and mean minus one standard deviation of each random variable in training and testing of the ANN model. Since two variables are used (frequency and mode shape) in this study, two upper limits $(\alpha_{++}, \alpha_{-+})$ and two lower limits $(\alpha_{--}, \alpha_{+-})$ need to be obtained. Thus, four ANN models are developed by considering the mean values and standard deviations (σ) of the random noises applied to each variable. The training functions of the four ANN models are listed in Table 14. The relationship between input and output parameters of ANN model is denoted by $fn(\cdot)$

Table 14. Training functions for ANN model

Model (n)	Training function
1	$\alpha_{j_{++}}^n = fn(\lambda_i^0 + \sigma_{\lambda_i}, \phi_i^0 + \sigma_{\phi_i})$
2	$\alpha_{j_{--}}^n = fn(\lambda_i^0 - \sigma_{\lambda_i}, \phi_i^0 - \sigma_{\phi_i})$
3	$\alpha_{j_{+-}}^n = fn(\lambda_i^0 + \sigma_{\lambda_i}, \phi_i^0 - \sigma_{\phi_i})$
4	$\alpha_{j_{-+}}^n = fn(\lambda_i^0 - \sigma_{\lambda_i}, \phi_i^0 + \sigma_{\phi_i})$

where $\alpha_{j_{++}}^n, \alpha_{j_{--}}^n, \alpha_{j_{+-}}^n, \alpha_{j_{-+}}^n$ are the target outputs of ANN models trained with different combinations of mean plus one standard deviation and mean minus one standard deviation of frequencies and mode shapes for the jth structural element, in this study the stiffness of the jth element; σ_{λ_i} and σ_{ϕ_i} are the standard deviation of the ith frequency and mode shape, and n is the ANN model number. Using the trained ANN model and the measured frequency and mode shape λ_i, ϕ_i as input, the stiffness of each element $\hat{\alpha}_{j_{++}}^n, \hat{\alpha}_{j_{--}}^n, \hat{\alpha}_{j_{+-}}^n, \hat{\alpha}_{j_{-+}}^n$ can be obtained for the nth ANN model. The means $E(\hat{\alpha})$ and standard deviations $\sigma(\hat{\alpha})$ are calculated by

$$E(\hat{\alpha}) = \frac{1}{16}(\hat{\alpha}^1_{j++} + \hat{\alpha}^1_{j+-} + \hat{\alpha}^1_{j-+} + \hat{\alpha}^1_{j--} + \ldots + \hat{\alpha}^4_{j++} + \hat{\alpha}^4_{j+-} + \hat{\alpha}^4_{j-+} + \hat{\alpha}^4_{j--})\ (14)$$

$$\sigma(\hat{\alpha}) = \left[E(\alpha^2) - (E(\alpha))^2 \right]^{1/2} \tag{15}$$

Once the means and standard deviations of element stiffness are obtained, the probability of damage existence (PDE) can be estimated from statistical distributions of the stiffness parameters of the undamaged and damaged models. The basic idea is to compute the probability of each stiffness parameter contained within a defined damage interval at a specific confidence level. For example, if the stiffness parameter (α_j) of the undamaged element j is normally distributed with mean $E(\alpha_j)$ and standard deviation $\sigma(\alpha_j)$, the probability density function can be obtained as illustrated in Figure 19 , where L_{α_j} is the lower bound of the healthy parameter. By setting a 95% confidence level, the lower bound is $L_{\alpha_j} = E(\alpha_j) - 1.645\sigma(\alpha_j)$, which indicates that there is a probability of 95% that the healthy stiffness parameter falls in the range of $[E(\alpha_j) - 1.645\sigma(\alpha_j), \infty]$. Similarly, for the stiffness parameter of element j in the damaged state (α'_j), the distribution is again assumed as normal with mean $E(\alpha'_j)$ and standard deviation $\sigma(\alpha'_j)$, and the corresponding probability density function is also plotted in Figure 19. The PDE is defined as the probability of α'_j outside the 95% confidence healthy interval. Thus the PDE of element j is

$$P^j_d = 1 - prob(L_{\alpha_j} \le x_{\alpha'} \le \infty) = prob(-\infty \le x_{\alpha'} \le L_{\alpha'}) \tag{16}$$

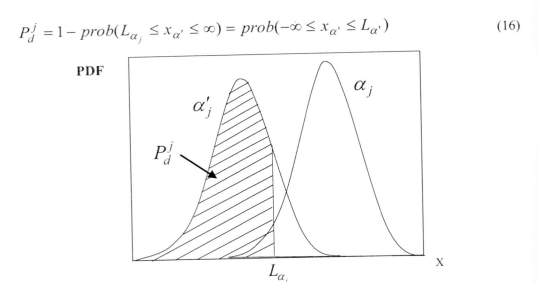

Figure 19. Probability density functions of α_j and α'_j and probability of damage existence P^j_d

PDE is a value between zero and one, and if the PDE of a segment is close to one, it is most likely the element is damaged. If the PDE is close to zero, damage existing in the element is very unlikely. It should be noted that numerically simulate results, which are not shown here, proved that the stiffness parameters of the undamaged and damaged state indeed have normal distributions if the random variations of modeling errors and measurement noises are zero mean normally distributed random variables.

To demonstrate the ANN model with consideration of both modeling error and measurement noise in identifying structural damage, the laboratory tested concrete slab presented above is used as an example. The concrete slab is discretized to 52 elements with 81 nodes. To reduce the computational effort for training the ANN model, these elements are lumped to form seven segments as shown in Figure 20. Stiffness in each segment is assumed the same in calculations. A total of 1200 cases with stiffness in each segment varying from 0.2 to 1.5 times of the true stiffness are simulated to train and test the model. Details of the ANN model and training and testing can be found in Bakhary et al. (2007).

First an ANN model is trained without considering modelling errors and measurement noise. Figure 21 demonstrates the accuracy of this ANN model in detecting both the location and severity of single and multiple damages in the slab by using the noise-free numerical data, in which SRF is defined as the stiffness reduction of the segment. The performance of this model, however, is not as good when the actual measured data is used to detect damage in the tested RC slab. Figure 22 shows the damage predictions by the trained ANN model at some loading levels. As shown, the ANN model fails to yield correct damage predictions. For example, in loading levels 3 and 4, the damage is limited to the left span, mainly in segment 2 as shown in Figure 10, but the predicted largest damage occurring in segments 3 and 6. Moreover, at loading levels 11 and 12, as shown in Figure 10, damage spreads almost to the entire slab, but the predicted damage at loading level 11 is only in segment 6. At loading level 12, although wide spread damage in the slab is predicted, the model fails to predict damage in segment 1 and 7 associated with support movements. These observations indicate the trained ANN model based on noise-free numerical data cannot reliably predict the actual RC slab damage by using measured vibration properties because they are contaminated with noises.

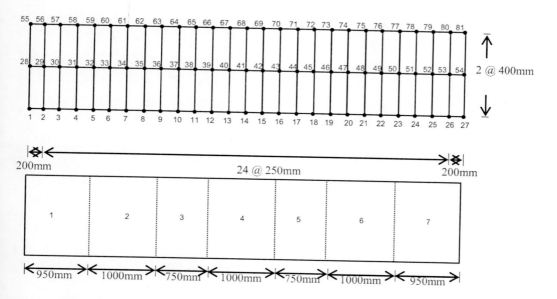

Figure 20. FE model of the tested RC slab

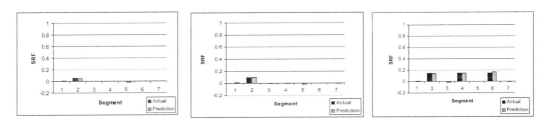

Figure 21. Detection of numerically simulated damage by ANN model

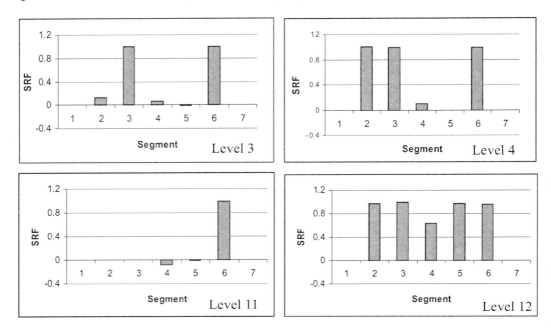

Figure 22. ANN model prediction of damage in tested RC slab

Following the procedure described above by assuming a 2% coefficient of variation (COV) in frequencies and 15% COV in mode shapes of the FE model simulated data for training the ANN model and assume the same levels of COV in the measured frequencies and mode shapes, the trained ANN model is used to predict damage in the tested RC slab. Figure 23 shows the calculated PDE of each segment at each loading level. From the figures, it is seen that the PDEs calculated using statistical ANN agreed with the crack propagation trend in the test observations shown in Figure 10. The low PDEs predicted at level 1 and level 2 at all segments indicate that there is no significant crack detected. This agrees with the observations in the experiment when 6KN and 12KN loads were applied to the left span. At level 3, the highest PDE value is obtained at segment 2 (middle of left span), which is the observed damage location.

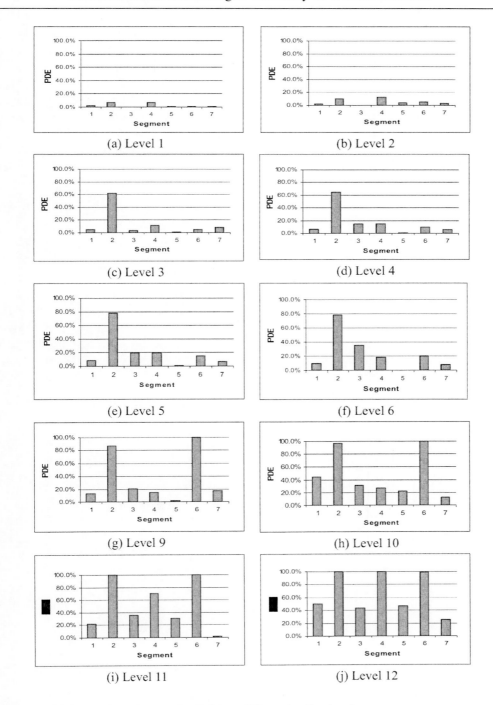

Figure 23. Predicted PDE of the tested RC slab at different loading levels

At loading levels 4 to level 6, the PDEs at segment 2 remain about the same, but the PDEs at segments 3, 4, 6 increase with loading level, indicating that there are no significant changes at the middle of left span but small damages occurred at the other segments. These results also agree with the observations in the test, At loading level 9, the highest PDEs occurred at segments 2 and 6 (middle of the left and right span), indicating that the damage observed in the experiment at the both spans are correctly identified. The PDEs pattern for levels 10, 11 and 12 also show a good agreement with the observed crack propagation, with the PDEs at segments 2 and 6 increased to 100% at level 12. The PDE of segment 4 also shows the similar trend where it increases from 27.8% at level 8 to 100% at level 12, identifying the damage at the middle support. The PDEs at other segment also show an incremental trend as the load level increases. The high PDEs at segments 1 and 7 at loading level 12 are caused because of the support movement as mentioned above. These results demonstrate that the predicted PDEs are close to the damage pattern observed in the test, indicating that considering modeling errors and measurement noise in SHM calculations will provide better predictions of structural damage.

CONCLUSION

Finite element modeling error and structural vibration measurement noise are unavoidable. If they are not properly considered, they may lead to erroneous SHM. This chapter presented some research results obtained in UWA in the last few years by studying the finite element modeling errors and measurement noises for a more reliable SHM. An ANN model that considers both modeling error and measurement noise for SHM was also introduced. Its accuracy was proved by testing data. It was demonstrated that considering these uncertainties leads to better structural condition identifications.

ACKNOWLEDGEMENT

This chapter summarizes some of the research results related to SHM obtained in the School of Civil and Resource Engineering, the University of Western Australia. Many people contributed to the research work, including colleagues in the Structural Engineering group, Prof. Andrew Deeks, Dr. Yong Xia, Dr. Giovanna Zarnado, Dr. Xinqun Zhu, PhD students Norhisham Bakahry and Xueling Peng, and many undergraduate students and technicians in the Structures Lab. Financial support from ARC, CIEAM and Main Roads WA are also acknowledged.

REFERENCES

Askegaard V, Mossing P. (1998). "Long-term Observations of RC-Bridge Using Changes in Natural Frequency." *Nordic Concrete Research*, 7, pp20-27.
Bayissa, WL and Haritos, N. (2005a). "Structural damage assessment using mean-square value of response power spectral density." *Proceedings of Australian Structural Engineering Conference*, Newcastle, Australia, CD-Rom.
Bayissa, WL and Haritos, N. (2005b). "Structural damage identification using spectral moments of the vibration response signal." *Proceedings of 1st International Conference on Structural Condition Assessment, Monitoring and Improvement*, Perth, Australia.

Brincker, R, Zhang, L and Andersen, P. (2000). "Modal identification from ambient responses using frequency domain decomposition." *Proc. the 18th IMAC*, San Antonio, USA. pp625-630

Bakhary, N, Hao, H, and Deeks (2007). "A. Damage detection using artificial neural network with consideration of uncertainties." *Journal of Engineering Structures*. 29, pp2806-2815.

Cawley P. (1997). "Long Range Inspection of Structures Using Low Frequency Ultrasound: Structural Damage Assessment Using Advanced Signal Processing Procedures." *Proceedings of DAMAS 97 (Euromech 365)*. Sheffield Academic Press, UK, pp1–17.

Cioara TG, Alampalli S. (2000). "Extracting Reliable Modal Parameters for Monitoring Large Structures." *Proceedings of European COST F3 Conference on System Identification and Structural Health Monitoring*. Madrid, Spain, pp333–340.

Cornwell P, Farrar CR, Doebling SW, Sohn H. (1999). "Environmental Variability of Modal Properties." *Experimental Techniques*, 23(6), pp45-48.

Doebling, SW, Farrar, CR, Prime, B P & Shevitz, DW. (1996). "Damage Identification and Health Monitoring of Structural and Mechanical Systems from Changes in Their Vibration Characteristics: A Literature Review." *LA-13070-MS, Los Alamos National Laboratory*, Los Alamos, New Mexico.

Doebling, SW, Farrar, CR and Cornwell, P (1997). "A statistical comparison of impact and ambient testing results from the Alamosa Canyon Bridge." *Proc. 15th IMAC*, Florida, USA, pp264-270.

Ewins DJ. (2000). "Modal Testing – Theory, Practice and Application." *Research Studies Press Ltd*: Baldock, Hertfordshire, UK.

Farrar CR, Doebling SW, Cornwell PJ, Straser EG. (1997). "Variability of Modal Parameters Measured on the Alamosa Canyon Bridge." *Proceedings of the 15th International Modal Analysis Conference*. Orlando, FL, February 3-6, USA, pp257-263.

Farrar, CR and Doebling, SW (1998). "A comparison study of modal parameter confidence intervals computed using the Monte Carlo and Bootstrap techniques." *Proceedings of the 16th International Modal Analysis Conference*. Feb, Santa Barbara, CA, USA, pp936-944.

Farrar, CR & Jauregui, D (1996), "Damage Detection Algorithms Applied to Experimental and Numerical Modal Data from the I-40 Bridge." *LA-13074-MS, Los Alamos National Laboratory*, Los Alamos, New Mexico.

Khahil A, Greimann L, Wipf T, Wood D. (1998). "Modal Testing for Nondestructive Evaluation of Bridges: Issues." *Crossroads 2000 Proceedings*. Ames, Iowa, August 19-20, pp109-112.

Liberatore, S & Carman, GP (2004). "Power spectral density analysis for damage identification and location." *Journal of Sound and Vibration*, 274, pp761-776.

Liu, PL (1995). "Identification and damage detection of trusses using modal data." *Journal of Structural Engineering, ASCE,* 121(4), pp559-608.

Low, HY and Hao, H (2002). "Reliability analysis of direct shear and flexural failure modes of RC slabs under explosive loading." *International Journal of Engineering Structures*, 24, pp189-198.

Maeck J, Peeters B, De Roeck G. (2000). "Damage identification on the Z24-bridge using vibration monitoring analysis." *Proceedings of European COST F3 Conference on System Identification and Structural Health Monitoring*. Madrid, Spain, June 6-9 pp233-242.

Malhotra, V. (ed.) (1984), "In Situ Nondestructive Testing of Concrete." *American Concrete Institute*, Michigan, USA

Modena, C, Sonda, D & Zonta, D. (1999). "Damage localization in reinforced concrete structures by using damping measurements." *Proceedings of International Conference on Damage Assessment of Structures*, Dublin, Ireland, pp132-141.

Montalvão, D, Maia, NMM & Ribeiro, AMR. (2006). "A review of vibration-based structural health monitoring with special emphasis on composite materials." *The Shock and Vibration Digest*, 38, pp295-324.

Nashif, AD, Jones, DIG and Henderson, JP. (1985). *Vibration Damping*. New York: John Wiley & Sons.

Ndambi, J-M, Vantomme, J & Harri, K. (2002). "Damage assessment in reinforced concrete beams using eigen frequencies and mode shape derivatives." *Engineering Structures*, 24, pp501-515.

Pandey, AK, Biswas, M & Samman, MM. (1991). "Damage detection from changes in curvature mode shapes." *Journal of Sound and Vibration*, 145, pp321-332.

Papadopoulos, L and Garcia, E. (1998). "Structural damage identification: A probabilistic approach." *AIAA Journal*, 36(11), pp2137-2145.

Pessiki, S and Olson, L. (ed), (1997). "Innovations in Non-Destructive Testing of Concrete." *American Concrete Institute*, Michigan, USA

Peeters B, De Roeck G. (2001). "One-year monitoring of the Z24-Bridge: environmental effects versus damage events." *Earthquake Engineering and Structural Dynamics*, 30, pp149–171.

Peeters B, Maeck J, De Roeck G. (2000). "Vibration–Based Damage Detection in Civil Engineering: Excitation Sources and Temperature Effects." *Smart Materials and Structures*; 10, pp518-527.

Ricles, JM and Kosmatka, JB. (1992). "Damage detection in elastic structures using vibratory residual forces and weighted sensitivity." *AIAA Journal*. 30(7), pp2310-2316.

Rohrmann RG, Baessler M, Said S, Schmid W, Ruecker WF. (2000). "Structural Causes of Temperature Affected Modal Data of Civil Structures Obtained by Long Time Monitoring." *Proceedings of the 18th International Modal Analysis Conference*. Kissimmee, Florida, pp1–7.

Rohrmann RG, Rucker WF. (1994). "Surveillance of Structural Properties of Large Bridges using Dynamic Methods." *Proceedings of 6th International Conference on Structural Safety and Reliability*. Innsbruck, Austria, August 9-13; pp977-980.

Rosenbleuth, E, (1975). "Point estimates for probability moments." *Proc. National Academy of Science,* 72(10), pp3812-3814.

Rucker WF, Said S, Rohrmann RG, Schmid W. (1995), "Load and condition monitoring of a highway bridge in a continuous manner." *Proceedings of the IABSE Symposium on Extending the Lifespan of Structures*. San Francisco, CA, Vol 73.

Salawu, OS (1997). "Detection of structural damage through changes in frequency: a review." *Engineering Structures*, 19, pp718-723.

Salawu, OS & Williams, C. (1995). "Bridge assessment using forced-vibration testing." *Journal of Structural Engineering, ASCE*, 121, pp161-173.

Salzmann. A, Fragomeni, S and Loo, YC. (2002). "The Damping Analysis of Experimental Concrete Beams under Free Vibration." *Advances in Structural Engineering*, vol 6, No. 1, pp.53-63.

Sohn H, Dzwonczyk M, Straser EG, Kiremidjian AS, Law KH, Meng T. (1999). "An Experimental Study of Temperature Effect on Modal Parameters of the Alamosa Canyon Bridge." *Earthquake Engineering and Structural Dynamics*, 28 (8), pp879–897.

Sohn, H, Farrar, CR, Hemez, FM, Shunk, DD, Stinemates, DW & Nadler, BR. (2003). "A Review of Structural Health Monitoring Literature: 1996-2001." *LA-13976-MS, Los Alamos National Laboratory*, Los Alamos, USA.

Van, OP and de Moor, B. (1996). *Subspace identification for linear system: Theory, implementation and Applications.* Kluwer Academic Publishers, Dordrecht, Netherlands.

Wang Y, Zhu XQ, Hao H and Ou JP. (2009). "Guided wave propagation and spectral element method for debonding damage assessment in RC structures." *Journal of Sound and Vibration*, 324, pp751-772.

Xia Y, Hao H. (2003). "Statistical Damage Identification of Structures with Frequency Changes." *Journal of Sound and Vibration*, 263 (4), pp853-870.

Xia, Y, Hao, H, Zanardo, G & Deeks, A. (2006) "Long-term vibration monitoring of an RC slab: Temperature and humidity effect." *Engineering Structures*, 28, pp441-452.

Xia, Y, Ren, WX, Hao, H, Zhu, XQ and Bakhary, N. (2007). "Impact and ambient vibration testing of a new highway bridge." *Proc. The Int. Symposium on Integrated Life-Cycle Design and Management of Infrastructures*, Shanghai, China.

Zhao, J & DeWolf, JT. (1999). "Sensitivity study for vibrational parameters used in damage detection." *Journal of Structural Engineering, ASCE*, 125, pp410-416.

In: Structural Health Monitoring in Australia
Editors: Tommy H.T. Chan and D. P. Thambiratnam

ISBN: 978-1-61728-860-9
©2011 Nova Science Publishers, Inc.

Chapter 5

VIBRATION-BASED DAMAGE DETECTION FOR TIMBER STRUCTURES IN AUSTRALIA

B. Samali[*], J. Li[±], U. Dackermann and F.C. Choi
University of Technology Sydney, Australia

ABSTRACT

The use of non-destructive assessment techniques for evaluating structural conditions of aging infrastructure, such as timber bridges, utility poles and buildings, for the past 20 years has faced increasing challenges as a result of poor maintenance and inadequate funding. Replacement of structures, such as an old bridge, is neither viable nor sustainable in many circumstances. Hence, there is an urgent need to develop and utilize state-of-the-art techniques to assess and evaluate the "health state" of existing infrastructure and to be able to understand and quantify the effects of degradation with regard to public safety. This paper presents an overview of research work carried out by the authors in developing and implementing several vibration methods for evaluation of damage in timber bridges and utility poles. The technique of detecting damage involved the use of vibration methods, namely damage index method, which also incorporated artificial neural networks for timber bridges and time-based non-destructive evaluation (NDE) methods for timber utility poles. The projects involved successful numerical modeling and good experimental validation for the proposed vibration methods to detect damage for simple beams subjected to single and multiple damage scenarios and was then extended to a scaled timber bridge constructed under laboratory conditions. The time-based NDE methods also showed promising trends for detecting the embedded depth and condition of timber utility poles in early stages of that research.

---

[*] Centre for Built Infrastructure Research (CBIR), University of Technology Sydney, Australia (Bijan.Samali@uts.edu.au

[±] Centre for Built Infrastructure Research (CBIR), University of Technology Sydney, Australia (Jianchun.Li@uts.edu.au

INTRODUCTION

Australia has been, historically, blessed with plentiful supplies of native hardwood timbers. Such timber is of both strong and durable species. As a result, these materials were heavily used during the early European settlement since 1788, for developing transport links such as wharf and bridge structures. Later, timber was used for utility poles for transmission of electricity and communication lines (Crews, 2008).

Recent surveys have shown that there are approximately 29,000 timber bridges in Australia, of which many are still in service despite being degraded or in structurally weakened condition (DoTaRS, 2003). The enormous number of timber bridges, which form a significant portion of bridges in Australia (total is in excess of 40,000), reflects the importance of these structures to the country's economic growth and the daily life of many Australians. In addition, these bridges are highly valued, not just for economic reasons, but also for the social and historical value placed on them by rural communities.

The Department of Transport and Regional Services Australia (DoTaRS, 2003) has estimated that a third of timber bridges are in excess of 50 years old. These bridges are located mainly on regional roads, owned and maintained by local government agencies. Due to lack of funds and engineering expertise, it is not easy to undertake thorough condition assessment of these bridges, which are often functionally obsolete and structurally deficient.

Due to their age and unknown "health" condition, it is extremely crucial for asset managers to have a reliable tool for monitoring and evaluation of the state of health of their bridge stocks in order to carry out timely maintenance and repair/replacement, and therefore ensure the safety of public as well as preserve these heritage-valued structures for the country (McInnes, 2005).

Another important component of Australia's infrastructure is timber utility poles. According to Nguyen et al. (2004) and Francis and Narton (2006), there are nearly seven million utility poles in the current network, of which around five million timber poles are used for distribution of power and communications. The utility pole industry in Australia spends about \$40~50 million annually on maintenance and asset management to avoid failure of the utility lines, which is very costly and may cause serious consequences. Each year, about 30,000 electricity poles are replaced in the eastern states of Australia, despite the fact that up to 80% of these poles are still in a very good serviceable condition. For timber utility poles, investigation of asset management systems for the power distribution industry shows that the design and assessment methods, which form traditional industry practices in Australia, are imprecise and often unreliable. Assessment techniques are required to maintain a high level of reliability against pole failure. There is a need for a simple, reliable and effective tool for determination of embedded length and condition of timber utility poles in service.

This chapter reports on research work carried out by the authors in development and implementation of vibration-based methods for evaluation of damage and state of health of timber structures. For damage detection and damage severity estimation, a modified damage index method (MDI) has been developed for dealing with timber beam-like structures, and a damage index method for plate-like structures (DI-P) was proposed for timber bridges and timber decks. Both the methods were formulated based on modal strain energy. In addition to numerical investigations using finite element models, the researchers carried out experimental studies on timber beams and a laboratory prototype timber bridge subjected to single- and multiple-damage scenarios. To enhance the damage detection results and to overcome shortcomings of the proposed modal-strain-energy-based methods, integration of artificial neural networks (ANNs) in form of a network ensemble was also proposed in damage localization and severity estimation. The research and development of damage detection in timber utility poles including use of non-destructive evaluation (NDE) techniques for determination of condition and underground depth of embedded timber poles in-service is also briefly reported in this chapter.

LITERATURE REVIEW

State-of-the-Art in Damage Detection for Timber Structures

Timber/wood has been used extensively as building material for centuries. The degradation of structural members of timber may be due to several causes, such as biotic degradation and physical degradation. It is, therefore, important to perform periodical condition assessment and evaluation on the structures to determine the extent of deterioration for a proper maintenance schedule. Often the problems in wood/timber structures do not lie in the effectiveness of preserving systems of the material itself but in not having the correct tools necessary to locate deterioration (Duwadi et al., 2000). Thereby, it delays application of remedial treatments to prevent further degradation in the structures. The current state-of-the-art for in-situ condition assessment and evaluation of timber consists of several steps either through non-destructive, destructive or semi-destructive means as briefly listed in Table 1 (Kasal, 2008). A wide variety of techniques being used to verify deterioration in timber members can be found in Ross et al. (2006).

Table 1. Classification Of Methods For In-Situ Timber Evaluation (Kasal, 2008)

Method		
NON-DESTRUCTIVE	DESTRUCTIVE	SEMI-DESTRUCTIVE
Visual inspection	Specimen extraction	Resistance drilling
Stress and acoustic waves	Full-member tests	Core drilling
Moisture contentmeasurements	Standard tests of mechanical properties	Tension micro-specimens
Radiography		Pin penetration resistance
Computer tomography		Other non-conventional methods
Species determination		
Dendrochronology		

It is quite obvious that destructive tests impose several disadvantages such as complete loss of the element of interest. For semi-destructive methods, they are non-destructive with respect to the member of interest, but are destructive with respect to the extracted specimen itself. Hence, non-destructive methods are widely used by engineering practitioners lately, as these methods do not alter much the tested member material properties. However, many non-destructive techniques (NDTs) are mostly used on small and accessible areas. It is time consuming to perform them on large timber structures like timber bridges, and it is also costly. This led to research on global methods, which would be able to reduce testing time, the cost involved and disruption to traffic flow. If necessary, the global methods can be combined with other more localized NDE methods for evaluating damage details.

Over the last few decades, many methods assessing global structural integrity for timber structures have been developed. Some recent works have focused on using vibration techniques, which are non-destructive in nature, to assess the in-service stiffness of timber structures like short-span timber bridges and floor systems within buildings. For timber structures, the focus of developments on vibration techniques has been on assessing the condition of wood in-service based on techniques that evaluate individual members and small area defects within a wood member. The basis of vibration-based methods is that any changes in the physical properties of wood such as mass, damping and stiffness, will in turn alter its dynamic characteristics, namely, natural frequency, damping ratio and mode shape.

Peterson et al. (2001a,b) reported on adopting the damage index method for one-dimensional (1-D) systems that utilizes modal strain energy before and after damage for identifying local damage and decay in timber beams using the first two flexural modes. The researchers also mentioned that the mode shapes to be used in the damage localization algorithm should be selected according to the location of damage. The method is quite promising for single-damage scenarios but it does not work well for two-damage scenarios in a timber beam. Furthermore, the results showed that the method is quite mode dependant.

Morison et al. (2002) presented the possibility of extending global testing methods, specifically, impact testing, for determining if damage in the form of decay is present in timber bridges. The project also intended to provide cost and time savings over existing methods of inspecting timber bridges. The method is to correlate the first natural frequency (fundamental frequency) with stiffness characteristics of timber bridge stringers (girders) using an algorithm derived from a partial differential equation governing the transverse vibration of a simple flexural beam. However, the largest obstacles encountered in this study were the actual boundary conditions of a bridge in the field, which are difficult to determine. Similar efforts were extended by Morison et al. (2003) in investigating dynamic characteristics of timber bridges using several existing methods such as damage index derived from estimated and measured curvature and flexibility influence coefficient estimated by the direct current (DC) value of the frequency response function (FRF) matrix. The results showed that the broadband curvature method was successful in showing a difference between the original state and the damaged state. However, it is far from a quantitative prediction of damage. While the flexibility influence coefficient method was able to show a change in the system when a weaker stringer replaces an original one, it failed to identify which stringer was replaced. Similar work was reported by Wang et al. (2005).

Hu and Afzal (2006) used a damage indicator for the detection of single- and two-damage cases of timber beams. The damage indicator is derived from the difference of the mode shapes before and after damage. The indicator is then expressed using discrete Laplace transform and finally normalized statistically in space. The method using either first or second modes successfully identified single artificial defect at different locations and a two-damage case with equal spacing and similar level of damage severity. However, the method is incapable of evaluating quantitatively the severity of damage. In addition, the method found difficulty in generating higher mode shapes as well as the combination of first and second modes. The damage indicator was further sensitive to different damage locations and different modal orders. The results of this study were based on 43 impact points for each set of data from the modal tests, which may be too time consuming and, hence, not cost effective.

Muller (2008) presented a paper describing the application of ground penetrating radar (GPR) technique to detect defects within the timber superstructure of the Hornibrook Highway Bridge. The author found that GPR was successful in detecting, locating and estimating the extent and severity of girder defects on the bridge in around a fifth of the time and cost of the traditional approach of "visual inspection plus test drilling." Furthermore, the ability of GPR to locate isolated defects led to numerous defects being found that would have gone undetected based on the traditional visual inspection or targeted test drilling.

Damage Detection utilizing Artificial Neural Network Techniques

In recent years, the use of Artificial Neural Networks (ANNs) in structural damage detection has gained momentum. ANNs are artificial intelligence that simulate the operation of the human brain. They are composed of a large number of highly interconnected processing elements, which are analogous to neurons and are tied together with weighted connections that represent synapses. The key element of ANNs is that they learn by example and not by following programming rules. Once trained, they are capable of pattern recognition and classification and are robust in the presence of noise. These characteristics make ANNs powerful complementary tools in vibrational damage identification. With the use of ANNs, some critical issues in traditional damage detection methods can be overcome and damage detection accuracy and reliability greatly improved.

The first researchers to introduce ANN to vibration-based damage detection in civil structures were Wu et al. (1992). Their paper from 1992, investigates the feasibility of ANN in structural damage detection; Fourier spectrum recordings of an experimental three-storey frame structure were used as ANN inputs to detect damage. The researchers found that the network was successful in identifying the damage condition of each member in the structure and concluded that "the use of neural networks for structural damage assessment is a promising line of research."

Sahin and Shenoi (2002) used changes in natural frequencies and curvature mode shapes as input features for ANNs for location and severity prediction of numerical and experimental damage in cantilever steel beams. From the network predictions, they found that the reduction in natural frequency provides the necessary information for the existence and severity of damage, however, differences in curvature mode shapes severed as better indicator in the location predictions.

Many successful applications have also been reported on multi-stage damage assessment methods that incorporate artificial intelligence techniques. Ko et al. (2002) developed a three-stage identification scheme, which was applied to numerical simulations of the Kap Shui Mun cable-stayed bridge in Hong Kong. In the first stage, auto associative neural networks are fed with natural frequencies to identify the existence of damage. This was followed by a combination of modal curvature index and modal flexibility index to identify the damage area. In the third stage, ANNs were used to determine specific damage member(s) and damage severity.

Lee and Yun (2006) presented a two-step damage identification strategy and demonstrated the method on numerical data and field test data of the old Hannam Grand Bridge in Seoul, Korea. First, the DI method was used to screen potentially damaged members and then, neural networks were trained with mode shape differences between before and after damage to assess the defects. They found that while the damage screening of the first step gave many false damage alarms, the damage assessment results using neural networks showed good estimates for all damage cases.

Another multi-stage damage assessment approach was proposed by Bakhary et al. (2007), who used substructure techniques together with a two-stage ANN model to detect single- and multiple-damage scenarios. The method was successfully demonstrated on numerical simulations of a two-span concrete slab.

An extension to ANNs are neural network ensembles (also referred to as committees or classifier ensembles), which are a hierarchy of networks that are trained independently for the same task and whose outcomes are fused at different stages by ensemble networks. These neural network committees were observed to perform better than the best network used in isolation (Perrone and Copper, 1993) and are currently used in many fields of research.

The idea of multi-stage network training was first employed in the area of vibration-based damage detection by Marwala and Hunt (1999). The researchers applied a two-stage neural network ensemble to numerically simulated cantilever beam data, with one network trained with frequency energies, which are integrals of the real and imaginary components of the frequency response functions over various frequency ranges, and another network trained by using the first five flexural mode shape vectors. The authors found that the ensemble gave a mean error of 7.7% compared to 9.5% and 9.78%, respectively, of the individual networks. In an extension to this work, Marwala (2000) presented a committee of neural network approach, which employs frequency response functions (FRFs), modal properties (natural frequencies and mode shapes), and wavelet transform (WT) data of numerical and experimental steel seam-welded cylindrical shells. The data were separately fed to three neural networks, which were later combined in a committee network. It was found that the committee identified the damage cases better than the three approaches used individually.

Vibration-Based Assessment Work Undertaken at the Centre for Built Infrastructure Research at UTS

In the authors' research group (CBIR), studies on vibration-based assessment methods for timber structures started in early 2000. A simple dynamic testing method, based on structural frequency shift (SFS), was developed by Samali et al. (2002, 2003a,b) (related work is also published in Li et al., 2004; Crews et al., 2004; and Benitez and Li, 2002) to evaluate the global stiffness and strength of in-service timber bridges. This dynamic method is generally proven to be easily performed and cost effective. In addition, the operators are not required to have high skills and can be easily trained on the job to perform the dynamic timber testing. The method has been successfully used to undertake field-testing of more than 600 timber bridge spans across New South Wales (NSW). The testing procedure involves the attachment of accelerometers underneath the bridge girders and exciting the bridge with a modal hammer with and without extra mass blocks at mid-span of the bridge. In order to refine the method and to further reduce cost and testing time as well as evaluate some of the practical issues during the field testing, a project was initiated to investigate the effects of using alternative mass forms, such as use of a trailer to replace currently used mass blocks, and adding mass at locations other than mid-span, on the bridge vibrations and subsequent integrity assessment (Li et al., 2005). The authors have progressively expanded the research work to also detect the location of damage and to evaluate damage severity. The modal-strain-energy-based damage index method for beam-like structures (DI) and its modified version (MDI), as well as damage index (DI) method for plate-like structures (DI-P) form the basis for the developed location and severity identification methods. For improved damage evaluation, artificial neural network techniques were introduced in the detection procedures. Indices of the DI method and compressed frequency response functions are used as input parameters for neural network training. Selected publications are listed in (Choi, 2007; Choi et al., 2007; Choi et al., 2008a,b; Li et al., 2007; Samali et al., 2009; Dackermann et al., 2008a,b, 2009). Furthermore, the group also started another research project using vibration-based methods to estimate embedded length in soil for timber utility poles as well as determining their structural conditions (Zad, 2009). Some of the research outcomes that deal with the identification of damage locations and severities on timber beams and timber bridge decks are presented in the following sections.

DAMAGE DETECTION ON TIMBER BEAM STRUCTURE

Timber Beam Structure

Pin-pin supported timber beams were chosen to verify the developed methods on a simple structure. The timber beams are intended to represent scaled girders of typical timber bridges in Australia. The scaling for the beams was based on dynamic similitude of the girders in terms of their first natural frequencies, which usually range from 5 to 20 Hz.

Laboratory Timber Beam

Experimental timber beams were set up and tested in the Structures Laboratory of the University of Technology Sydney (UTS). The beams were of treated radiata pine sawn timber, measuring nominal dimensions of 45 mm by 90 mm in cross-section with a span length of 4,500 mm. The modulus of elasticity (MOE) of the beams was about 12GPa obtained from four-point bending tests. The experimental set up is displayed in Figure 1(a).

(a)   (b)

Figure 1. (a) Experimental test set up and (b) medium-size damage (27 mm × 45 mm) of timber beam

Two laboratory timber beam structures were tested and inflicted with various types of damage. Single- and multiple-damage scenarios with different damage locations and severities were investigated. All damage scenarios consisted of a rectangular opening from the soffit of the beam, located at 1/4, mid-span (1/2), and 3/4 of the span length to simulate pockets of rot. Damage descriptions are as follows, L, M, and S are used to denote "light," "medium," and "severe" damage, respectively. All inflicted damage occupied 1% of the total span length (45 mm) and consisted of 10%, 30%, or 50% of the beam depth, designated as damage cases L, M, and S, respectively. The 10%, 30%, and 50% of the beam depth cut in cross section correspond to losses of sectional "I" (moment of inertia) of 27.1%, 65.7%, and 87.5%, respectively. The cases described in this chapter are listed in Table 2. A medium-size damage is displayed in Figure 1 (b).

Table 2. Dimensions of Damage Inflicted in Timber Beams

Damage Case	Damage Scenario	Location per 8th of span length	Length l (mm)	Depth h (mm)	% reduction of "I"
1	4L	4	45	9	27.1
2	4M	4	45	27	65.7
3	4S	4	45	45	87.5
4	4S6L	4, 6	45	45, 9	87.5, 27.1
5	4S6S	4, 6	45	All 45	All 87.5

To obtain the modal parameters of the beam, experimental modal testing and analysis was performed. In modal testing, the beams were excited by an impact hammer, and the acceleration responses were measured by nine equally spaced piezoelectric accelerometers. The time history signals of the hammer and the acceleration responses were first amplified by signal conditioners and then recorded by a data acquisition system. The sampling rate was set to 10,000 Hz for a frequency range of 5,000 Hz and 8,192 data points, thus giving a frequency resolution of 0.061 Hz. The time history data were then transformed into the frequency spectra using fast Fourier transform (FFT). By dividing the frequency spectra signals of the response data by the frequency spectra of the excitation data, the frequency response functions (FRFs) were obtained. The first five flexural modes of the beam were identified by performing modal analysis utilizing the leading software from LMS (LMS CADA-X). The corresponding mode shapes were mass normalized, and in order to enhance the quality and effectiveness of damage identification, the mode shape vectors were reconstructed from nine to 41 data points utilizing cubic spline interpolation techniques. The experimental modal testing set up and modal analysis procedures are shown in Figure 2.

Numerical Timber Beam Model

For damage severity estimation, artificial neural network techniques are employed. To generate patterns for the training of the networks, a numerical model of the timber beam structure is created using the finite element analysis package ANSYS. The dimensions and material properties of the pin-pin supported beam model comply with the specifications of the experimental beams. The cross-section is modeled with 20 elements across the height and four elements along the width. The longitudinal direction of the model is divided into 201 nodes and 200 elements. Seven different damage locations with spacings of 562.5 mm (1/8th of the span length) are considered. The damage locations ("1" to "7") are displayed in Figure 3(a). For each of these locations, five different damage severities, with cuts of 10%, 20%, 30%, 40% and 50% of the beam depth are investigated, generating a total of 35 different damage cases. All inflicted damage are 45 mm in length and 9 mm, 18 mm, 27 mm, 36 mm and 45 mm in height, respectively. This corresponds to 27.1%, 48.8%, 65.7%, 78.4% and 87.5% loss of I. Damage is modeled by rectangular openings from the soffit of the beam along the span length. A medium-size damage is depicted in Figure 3(b).

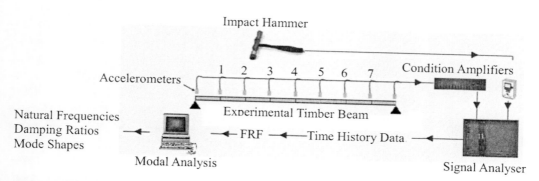

Figure 2. Schematic diagram of experimental modal testing and analysis

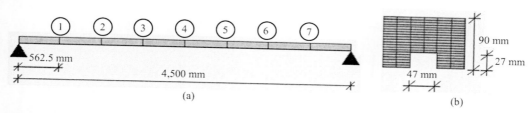

Figure 3. Finite element modeling of (a) pin-pin supported timber beam and (b) medium-size damage

Using the modal analysis module in ANSYS, the first five flexural mode shapes are extracted. To synchronise the numerically obtained data with the experimental data, the mode shape vectors are reduced from 201 data points to nine points, representing the nine measurement sensors of the experimental test set up. Subsequently, the mode shape vectors are again reconstructed from nine to 41 data points, utilizing cubic spline interpolation techniques.

Damage Localization Utilizing MDI Method

Derivation of MDI Method

The damage index (DI) method developed by Kim and Stubbs (1995) was adopted and modified (named as modified damage index (MDI)) to locate damage. The MDI method is based on relative differences in modal strain energy between an undamaged structure and that of the damaged structure. The modal strain energy utilizes mode shape curvature in the algorithm to calculate the damage index for the jth element and the ith mode, β_{ij}, as given below.

$$\beta_{ij} = \frac{\int_{j} \{\phi_i''^{*}(x)\}^2 \, dx \int_0^L \{\phi_i''(x)\}^2 \, dx}{\int_{j} \{\phi_i''(x)\}^2 \, dx \int_0^L \{\phi_i''^{*}(x)\}^2 \, dx} \tag{1}$$

In Eqn (1), the terms $\phi_i''(x)$ or $\phi_i''^{*}(x)$ are normalized mode shape curvature coordinates. The curvature is normalized with respect to the maximum value of the corresponding mode for each mode of a beam structure. The normalized curvature used in the MDI method is shown to perform better in damage localization compared to its original formulae, where curvature was not normalized (Kim and Stubbs, 1995). The asterisk denotes the damage cases. For the original damage index method, although mode shape vectors have been mass normalized, the mode shape curvatures used for the damage index calculation are not normalized. Values of mode shape curvature are dependant on the shapes of each individual mode shape. Instead of reflecting the changes in the curvature due to damage, the summation of non-normalized mode shape curvatures will distort the damage index in favor of higher modes, which results in false damage identifications. The modified damage index (MDI) method introduced above overcomes the problem by normalizing mode shape curvatures with respect to the maximum norm of each mode shape curvature. To account for all available modes, NM, the damage indicator value for a single element j is given as

$$\beta_j = \frac{\sum_{i=1}^{NM} Num_{ij}}{\sum_{i=1}^{NM} Denom_{ij}} \tag{2}$$

where Num_{ij} = numerator of β_{ij} and $Denom_{ij}$ = denominator of β_{ij} in Eq. (1), respectively. Transforming the damage indicator values into the standard normal space, the normalized damage index Z_j is obtained

$$Z_j = \frac{\beta_j - \mu_{\beta_j}}{\sigma_{\beta_j}} \tag{3}$$

where μ_{β_j} and σ_{β_j} are the mean and standard deviation of β_j values for all j elements, respectively.

The estimation of the damage severity for element j is expressed by equation

$$\alpha_j = 1 - \frac{1}{\beta_j} \tag{4}$$

with α_j being the severity estimator.

A judgment-based threshold value is selected and used to determine which of the jth elements are possibly damaged, which in real applications is left to the user to define based on what level of confidence is required for localization of damage within the structure.

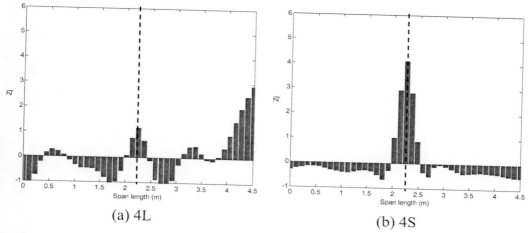

(a) 4L

(b) 4S

Figure 4. Damage localization for single-damage cases

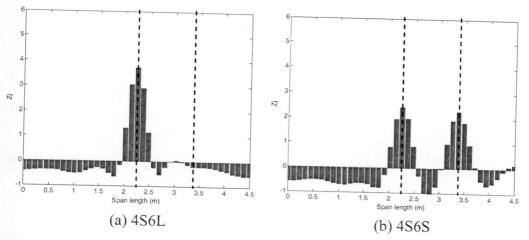

(a) 4S6L

(b) 4S6S

Figure 5. Damage localization for two-damage cases

Damage Localization Results

In the following, results of damage localizations utilizing the MDI method are presented. MDI indices Z_j are calculated using the first five flexural modes. In the results, the damage localization indices for each of the damage cases are plotted against the beam span length. In principle, any location with the index value Z_j larger than zero (the probability-based criterion for damage detection) is considered as damage existing at that location. The actual damage locations are indicated with dashed line in all figures. In Figure 4, for single-damage cases, the MDI is able to indicate accurately the location of damage at position 2.25 m, with few false positives (indication of false damage locations) at position 0.5625 m, position 3.375 m and near the supports for case 4L. Figure 5 depicts the damage scenarios of two damage locations positioned at 2.25 m and 3.375 m detected with MDI. The severe damage at mid-span (4S) is precisely identified by the damage indicator in both cases. This also applies to the damage at position 3.375 m in damage case 4S6S, however, the light damage of damage case 4S6L stays undetected. It is shown that the MDI method is capable of detecting severe damage, but may miss out light damage that appears together with severe damage. This is due to light damage altering slightly the mode shape and its derivatives, which may have been overshadowed by other more severe scenarios. It is observed that damage index Z_j does change accordingly with increase severity of damage as seen in Figure 4. However, damage index Z_j (indication of probability of damage existence at the location j) is not capable of estimating the severity of damage.

Damage Severity Estimation Utilizing DI Method and ANN

Methodology of Quantitative Prediction of Damage

To enhance the severity assessment of the DI method, neural network techniques are introduced to the damage detection algorithm. A hierarchy of neural networks, referred to as neural network ensemble, are designed to estimate the loss of the second moment of area (I) of single damage cases. The damage estimator α_j of the DI method is used as input parameter to the networks. The α_j values of various vibrational modes have different characteristics and are of varying importance. To respect these differences and to generate well-prepared input data sets, the individual neural networks are trained with mode separated damage estimators. The network ensemble is first trained with data generated from finite element models of the timber beam, and then damage indices of the laboratory timber beam are fed to the ensemble to test the networks and to predict the damage severities. Three experimental damage cases are investigated. The cases are all situated at mid-span of the beam and are of severities L, M and S (4L, 4M and 4S), which correspond to a loss of "I" of 27.1%, 65.7%, and 87.5%, respectively. The mid-span location was chosen, as this is a node point of mode 2 and mode 4 and, therefore, most problematic for the DI method.

Artificial Neural Network Architecture

A neural network ensemble is designed to estimate damage severities of timber beam structures. Numerical beam stimulations are used to train the networks and experimental timber beam data are used to test the networks with previously unseen data. First, five individual neural networks are separately fed with α_j values of the first five flexural modes, and then the outputs of each of the networks are combined in an ensemble network to produce the final severity estimation. All networks in the ensemble are multi-layer perceptron neural networks. The individual neural networks comprise an input layer with 41 nodes, corresponding to the 41 data points of the severity estimator α_j; three hidden layers with 30, 20 and 10 nodes and one single node output layer estimating the severity of the damage. The network ensemble is designed with five input nodes, which are the outputs of the five individual networks; three hidden layers of 7, 5, and 3 nodes and one output node predicting the damage extent. The transfer functions used are hyperbolic tangent sigmoid functions. Training is performed utilizing the back-propagation conjugate gradient descent algorithm. The inputs are divided into three sets: a training, a validation and a testing set. While the network adjusts its weights from the training samples, its performance is supervised utilizing the validation set to avoid overfitting. The network training stops when the error of the validation set increases while the error of the training set still decreases, which is the point when the generalization ability of the network is lost and overfitting occurs.

The complete data set of 35 numerical samples is allocated for training. For the laboratory beam data, the undamaged beam is experimentally tested five times, and the three damage cases are tested three times each. Thereby, a total of 45 α_j damage indices are generated (five undamaged data sets × three damaged data sets × three damage severity cases). The 45 experimental samples are divided into sets of 18 for validation and 27 for testing. The design and operation of all neural networks are performed with the software Alyuda NeuroIntelligence version 2.2 from Alyuda Research, Inc.

Damage Severity Results

In the following paragraph, the damage quantification predictions of the neural networks are presented. It is found that the results of the five individual networks greatly differ among each other. These prediction differences are a reflection of the individual characteristics of the different vibrational modes. In the subsequent figures, the horizontal-axis displays the 45 validation and testing samples of the three experimental damage cases 4L, 4M and 4S. The vertical-axis represents the normalized error, which is defined as $E_{norm}(d) = (T_d-O_d)/S_{max}$, where d is the damage case, T_d the target value, O_d the neural network predicted value and S_{max} the maximum severity (here, 100% loss of the second moment of area, "I"). Displayed in Figure 6 are the outcomes of the individual networks trained with the severity estimator α_j derived from (a) mode 2, (b) mode 3 and (c) mode 5. It is observed that the predictions of the network of mode 2 are incorrect for almost all damage cases. The reason for these errors is that for mode 2, the mid-span of the beam is a node point of flexural vibration and, therefore, the damage cannot be identified. For the network of mode 3, all medium and extra severe damage cases are below 10% normalized error; some large errors are obtained for extra-light damage. The network of mode 5 identifies all damage cases well with a maximum normalized error of 7.4%.

To determine the damage characteristics based only on the outcomes of the individual networks is problematic, as their damage predictions vary greatly. To achieve reliable damage identification, a conclusive, intelligent fusion of the network outcomes is necessary. This is achieved by the neural network ensemble, which combines the outcomes of the individual networks. Figure 7 displays the finial severity predictions of the ensemble network. It can be seen that the estimated severities of all damage cases eventually agree very well with the actual damage extents with a maximum normalized error of 4.4%. These results clearly show the effectiveness of the neural network ensemble approach, which produce results that are more accurate than any of the individual networks. These final predictions show that by introducing ANNs to the DI-based damage detection algorithm, shortcomings of the DI method, which was not able to quantify single light damage, can be overcome.

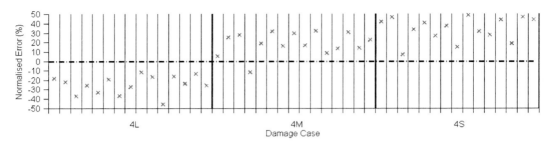

(a) Individual neural network outcomes trained with α_j derived from mode 2

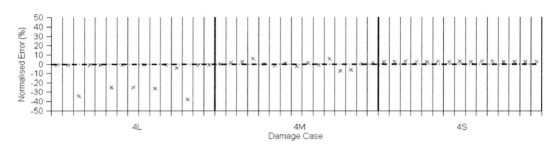

(b) Individual neural network outcomes trained with α_j derived from mode 3

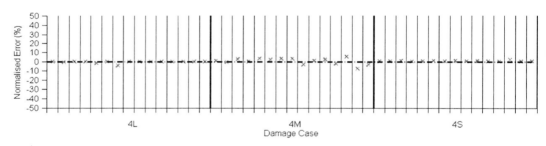

(c) Individual neural network outcomes trained with α_j derived from mode 5

Figure 6. Outcomes of individual neural networks tested with α_j damage indices derived from (a) mode 2, (b) mode 3 and (c) mode 5 to estimate the severity of experimental damage cases

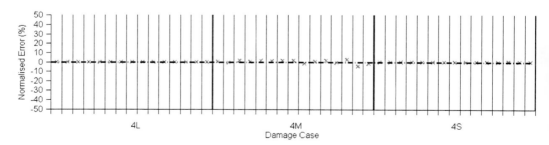

Figure 7. Final damage severity predictions of the neural network ensemble

DAMAGE DETECTION ON FOUR-GIRDER TIMBER BRIDGE

Four-Girder Laboratory Timber Bridge

A four-girder experimental timber bridge was built in the laboratory to verify the damage detection method for more complex timber structures. The basic dimensions of the deck structure are 2.4 m wide and 4.8 m long, with a span length of 4.5 m. The bridge was built of four girders (g1 to g4) of treated radiata pine sawn timber measuring 45 mm × 90 mm in cross-section. The deck consisted of four pieces of 21 mm thick × 2.4 m wide and 1.2 m long structural plywood of grade F11. The deck and girders were connected using 50 mm self-tapping screws with 137.5 mm spacing. No gluing was applied in order to avoid fully composite action, to simulate the interactions in a real timber bridge. The ends of the bridge girders were supported on concrete blocks and rigidly connected to the strong floor. A specially designed support system was used between girders and the concrete block to ensure a well-defined boundary condition that is very close to a pin-pin condition. The full bridge model is depicted in Figure 8. The goal of this study was to detect damage typically found in timber bridges. A total of 12 different damage scenarios at four damage locations were considered, as displayed in Figure 9. The inflicted damage cases consist of rectangular openings along the span of a girder starting from the soffit, located either at 1/8, 2/8, 4/8 (mid-span) or 6/8 of the span length to simulate pockets of rot. At each of these locations, three different damage severities were studied, denoted as light ("L"), medium ("M") and severe ("S") damage, respectively. All inflicted damage are 1% of the total span length (45 mm) and consist of 10%, 30% and 50% of the beam depth, corresponding to 27.1%, 65.7% and 87.5% loss of "I," respectively. The three damage severities are shown in Figure 10. The damage cases were inflicted in a cumulative manner. First, damage was inflicted on girder 2 (g2), then on girder 4 (g4), subsequently on girder 3 (g3) and finally on girder 1 (g1).

Figure 8. The laboratory timber bridge

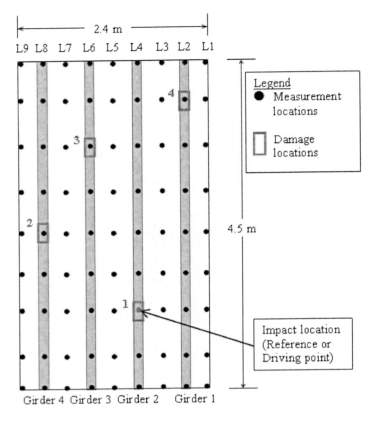

Figure 9. Plan view of damage and measurement locations on the bridge

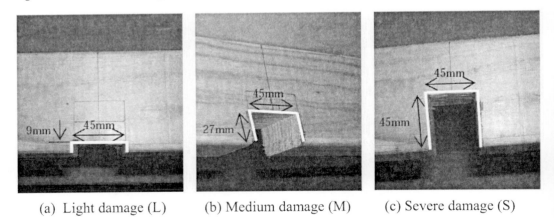

(a) Light damage (L)   (b) Medium damage (M)   (c) Severe damage (S)

Figure 10. Side view of various inflicted damage

Experimental Modal Analysis

The modal test set up and instrumentation layout is similar to the beam test as shown in Figure 2. However, the impact location (reference or driving point), which is situated at 3/4 of the span length and located directly above girder 2 (refer to Figure 9), is a strategic position that excites a large number of modes simultaneously. The response signal along the span was recorded using nine piezoelectric accelerometers. For the entire bridge deck, data from 81 measuring points (depicted in Figure 9) were acquired. Due to the limited number of available sensors, the data acquisition for the 81 points was done in 9 tests. The nine non-stationary accelerometers were moved from line-to-line, i.e., from line 1 (L1) to line 9 (L9), until all measurement locations were covered. These accelerometers were used to measure the acceleration response on the top surface of the bridge. The nine measurement points per line were deemed sufficient for accurate reconstruction of mode shapes using interpolation techniques. The driving point measurement enabled the experimental mode shapes to be mass normalized. The captured mode shapes are shown in Figure 11. From the 9×9-point experimental mode shapes, 41×41-point mode shapes were reconstructed using two-dimensional (2-D) cubic spline interpolation technique, generating mode shape vectors with 41 coordinates in both longitudinal and transverse directions. The reconstructed mode shapes with finer coordinates have shown to provide better chance of locating damage compared to using coarse coordinates. The reconstructed mode shapes were then applied in the damage detection algorithms attempting to locate and evaluate the inflicted damage scenarios.

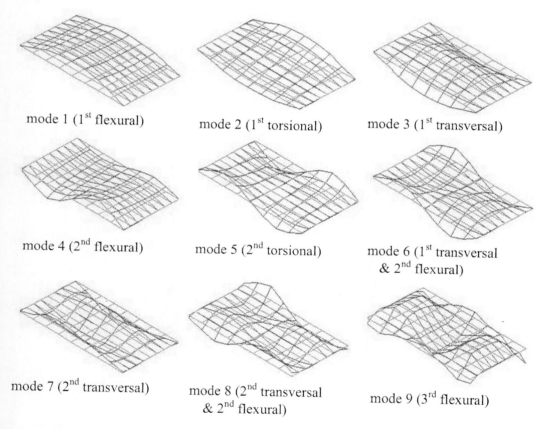

Figure 11. Experimental mode shapes (81-points) for the laboratory timber bridge

Damage Localization utilizing Damage Index for Plate-Like Structures (DI-P) for Bridge

Derivation of DI-P Method

Cornwell et al. (1999) extended the damage index method for beam-like structures to include the detection of damage for plate-like structures characterised by a two-dimensional (2-D) mode shape slope and curvature. The algorithm used to calculate the damage index for the jk^{th} subregion and the ith mode, β_{ijk}, is given below for detecting the location of damage.

$$\beta_{ijk} = \cfrac{\left\{ \int_{b_k}^{b_{k+1}} \int_{a_j}^{a_{j+1}} \left[\left(\frac{\partial^2 \phi_i^*}{\partial x^2}\right)^2 + \left(\frac{\partial^2 \phi_i^*}{\partial y^2}\right)^2 + 2v\left(\frac{\partial^2 \phi_i^*}{\partial x^2}\right)\left(\frac{\partial^2 \phi_i^*}{\partial y^2}\right) + 2(1-v)\left(\frac{\partial^2 \phi_i^*}{\partial x \partial y}\right)^2 \right] dxdy \right\}\left[\int_b^a \int \left(\frac{\partial^2 \phi_i}{\partial x^2}\right)^2 + \left(\frac{\partial^2 \phi_i}{\partial y^2}\right)^2 + 2v\left(\frac{\partial^2 \phi_i}{\partial x^2}\right)\left(\frac{\partial^2 \phi_i}{\partial y^2}\right) + 2(1-v)\left(\frac{\partial^2 \phi_i}{\partial x \partial y}\right)^2 dxdy \right]}{\left\{ \int_{b_k}^{b_{k+1}} \int_{a_j}^{a_{j+1}} \left[\left(\frac{\partial^2 \phi_i}{\partial x^2}\right)^2 + \left(\frac{\partial^2 \phi_i}{\partial y^2}\right)^2 + 2v\left(\frac{\partial^2 \phi_i}{\partial x^2}\right)\left(\frac{\partial^2 \phi_i}{\partial y^2}\right) + 2(1-v)\left(\frac{\partial^2 \phi_i}{\partial x \partial y}\right)^2 \right] dxdy \right\}\left[\int_b^a \int \left(\frac{\partial^2 \phi_i^*}{\partial x^2}\right)^2 + \left(\frac{\partial^2 \phi_i^*}{\partial y^2}\right)^2 + 2v\left(\frac{\partial^2 \phi_i^*}{\partial x^2}\right)\left(\frac{\partial^2 \phi_i^*}{\partial y^2}\right) + 2(1-v)\left(\frac{\partial^2 \phi_i^*}{\partial x \partial y}\right)^2 dxdy \right]}$$

$$(5)$$

where v represents Poisson's ratio, and the term $\partial^2 \phi_i / \partial x^2$ is a vector of second derivatives of mode shape coordinates (curvatures) with respect to x-axis. Similar convention is applicable to second derivatives with respect to y-axis and cross derivatives with respect to x- and y-axes. Equation (5) denotes damage index in matrix form describing change of strain energy before and after damage, corresponding to mode i for a bridge structure. The asterisk denotes the damage cases. To account for all available modes, NM, the damage indicator value for a single subregion jk is given as

$$\beta_{jk} = \frac{\sum_{i=1}^{NM} Num_{ijk}}{\sum_{i=1}^{NM} Denom_{ijk}}$$

$$(6)$$

where Num_{ijk} = numerator of β_{ijk} and $Denom_{ijk}$ = denominator of β_{ijk} in Eq. (5), respectively. Transforming the damage indicator values into the standard normal space, the normalized damage index Z_{jk} is obtained

$$Z_{jk} = \frac{\beta_{jk} - \mu_{\beta_{jk}}}{\sigma_{\beta_{jk}}}$$

$$(7)$$

where $\mu_{\beta_{jk}}$ = mean of β_{jk} values for all subregions jk and $\sigma_{\beta_{jk}}$ = standard deviation of β_{jk} for all subregions jk. A judgment-based threshold value is again selected and used to determine which of the subregions jk are possibly damaged.

Damage Localisation Results

In the following, the normalized damage indicator, Z_{jk}, calculated from the first nine modes is plotted against the laboratory timber bridge span length (4.5 m) and width (2.4 m). In principle, when the statistically normalized damage indicator value of Z_{jk} for a given location is larger than or equal to two (the probability-based criterion for damage), it is considered that damage exists at that location. To show the damage site in the contour plots, the actual damage locations are marked by an arrow sign and labeled with "Damage" in all subsequent figures.

The damage localization results of single damage cases are illustrated in Figure 12. For the light damage case g2L (the smallest damage), depicted in Figure 12(a), the method could not locate the damage at position (3.375 m, 0.9 m). For the severe damage case g2S, which is illustrated in Figure 12(b), the damage location was clearly detected, even though there were some false positives.

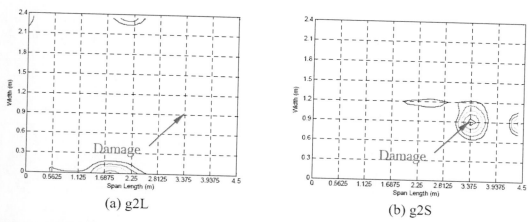

(a) g2L   (b) g2S

Figure 12. Single-damage cases using the experimental data

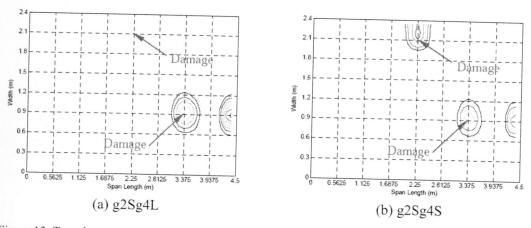

(a) g2Sg4L   (b) g2Sg4S

Figure 13. Two-damage cases using the experimental data

Observing from the results in Figure 13, severe damage locations at position (3.375 m, 0.9 m) were identified for damage cases g2Sg4L and g2Sg4S (see Figure 13(a) and (b), respectively) as well as severe damage at position (2.25 m, 2.1 m) for case g2Sg4S. However, the light damage at position (2.25 m, 2.1 m) was not identified for case g2Sg2L. A possible explanation is that the change in mode shape curvature caused by the severe damage is dominating the results and overwhelming the curvature change due to the light and medium damage. From the two damage cases, it should be noted that the second damage at position

(2.25 m, 2.1 m) was not detected until this damage became roughly comparable in severity to the first severe damage at position (3.375 m, 0.9 m) in the experimental studies. This indicates that the method may encounter problems in identifying multiple damage locations with different degrees of severity using experimental data. However, the method is capable of identifying all severe damage locations, and this is critical for a structure in terms of avoiding catastrophic failure. Detecting less severe damage in timber bridges using this method can be achieved as a two-stage process. Once the severe damage is identified and repaired, the method can be applied again and then any less severe damage can be detected in the absence of any severe damage. A detailed discussion on the assessment of all investigated timber bridge damage cases with the DI-P method can be found in Choi (2007).

Damage Severity Assessment Utilizing Frequency Response Functions and ANN

Methodology of Severity Detection

The DI-P method does not provide an algorithm for the estimation of damage severity. A different dynamic approach based on frequency response functions (FRFs) and artificial neural networks was developed to quantitatively estimate defects. FRF data are directly measured data, which have an abundance of information. They are easy to obtain and require very little human involvement (Fang and Tang, 2005). However, a very significant obstacle is the large size of the FRF data. Utilizing full-size FRFs in neural networks will result in a large number of input nodes, which will cause problems in training convergence and computational efficiency. If only partial sets of FRF data are used, an improper selection of data points from frequency windows will result in loss of important information, and errors will be introduced to the detection scheme (Ni et al., 2006). Principal Component Analysis (PCA) is a statistical technique for achieving dimensional data reduction and feature extraction. By projecting data onto the most important principal components, its size can greatly be reduced without significantly affecting the data. In the developed method, PCA is employed to compress the size of FRF data and make them suitable for network training. More precisely, residual FRFs, which are differences in FRF data between baseline and damage state, are compressed and used as ANN input. To demonstrate the method, it is applied to the different damage stages of the laboratory timber bridge. An accumulative damage growth, as simulated in the experimental bridge structure, is a common condition in field bridges. Unfortunately, in many circumstances, data on the intact state of these structures are not available. However, it is of importance for road engineers to obtain information on possible advancements of existing (even though unknown) damages as well as on newly acquired defects. To consider these needs, in this study, four different stages related to the four damage locations of the timber bridge are investigated. In the first stage, the undamaged structure is considered to be the baseline state, and the progress of damage at girder 2 is analyzed. In the second stage, the most severe damage case at girder 2 is regarded as baseline and the progress of damage at girder 4 is examined. In the third stage, the largest damage extent of damage at girder 4 (and the already existing damage at girder 2) is baseline and damage at girder 3 is investigated. Finally, the biggest damage state at girder 3 (including damage at girder 2 and 4) constitutes the baseline state and the extents of damage stages at girder 1 are analyzed. On each girder, nine accelerometer measurements are available. To minimize the effort for neural network training, the FRF data of the nine accelerometers of each girder are added up to obtain one summation FRF for each girder and each damage state. These summation FRFs are the bases for the network training.

Artificial Neural Network Architecture

Four neural networks are created utilizing the NeuroIntelligence software from Alyuda to estimate the severities of damage at girders 1 to 4. Each network is trained independently with data derived from summation FRFs of the four girders. First, differences in summation FRF data between the baseline state and the damage state are calculated. Then, PCA techniques are adopted to compress the residual FRFs onto its most important principal components. The PCA-compressed residual FRFs are then separated by girders and fed to the four neural networks to estimate the severity of damage. The created networks are back-propagation neural networks utilizing hyperbolic tangent sigmoid transfer functions and conjugate gradient descent training algorithms. The networks comprise of an input layer of nine nodes, representing the number of the selected PCs (the first nine principal components account for 97% of the original data); five hidden layers of 10, 8, 6, 4 and 2 nodes and one single node output layer estimating the extent of the damage. As each damage state is tested three times, a total of 108 data sets are available (three baseline data sets × three damaged data sets × four damage locations × three damage severities). The input set is again divided into a training, a validation and a testing set of 60, 24 and 24 data samples, respectively.

Damage Severity Outcomes

The damage severity estimations of the four neural networks are very promising. Each of the networks trained with measurements from one of the four bridge girders predicts the severity level of all 108 damage samples correctly. The outcomes of the network trained with PCA-compressed residual FRFs of girder 1 are displayed in Figure 14. These results show that damage fingerprints in FRF data are very good indicators for quantitatively estimating damage extents. As the predictions of each network give independently precise damage detections, a reduction of the sensor network is possible. Data from only nine sensors of one girder (or even less) are enough to successfully assess the severities of accumulative damage in the studied laboratory timber bridge.

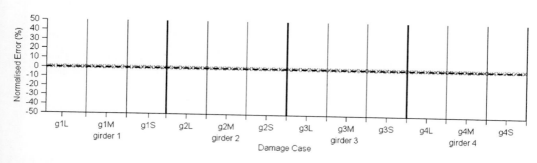

Figure 14. Outcome of neural network trained with PCA-compressed residual FRFs of girder 1

DETERMINATION OF EMBEDMENT DEPTH OF TIMBER POLE

Background of Research

Different types of non-destructive evaluation (NDE) techniques were developed during the last decades and used to evaluate the embedment depth and the quality of materials of embedded structures such as concrete piles. Some of these tests have also been utilized on timber piles or poles. However, the extent of knowledge developed on NDE techniques for timber poles and piles appears to be inadequate, but the effectiveness and reliability of the NDE techniques are not often questioned.

An ongoing research project of the Centre for Built Infrastructure Research (CBIR) at the University of Technology Sydney is the investigation on adequacy of the current NDE techniques for estimation of underground depth and condition of in-service timber pole structures as well as development of a new reliable and robust technique. The first stage of this research aims at conducting comprehensive state-of-the-art evaluations on available NDE techniques in terms of the reliability and effectiveness through numerical and experimental investigations. In the second stage, a new vibration- and acoustic-based NDE technique and procedure will be developed to encompass the latest development on advanced signal processing and damage detection, and to provide a reliable and effective tool for determination of the depth of embedment of timber poles/piles as well as their in-service condition including damage identification.

The current state-of-practice for non-destructive determination of unknown condition and embedded length of poles or piles are generally stress-wave-based and can be divided into two groups: the surface NDT methods and the borehole NDT methods. In the surface NDT, the reflection of the stress wave is measured directly from the pole/pile, while in the borehole methods, it needs a borehole drilled close to the pole/pile and extends along its length to perform tests. In this case, the reflection is monitored in the borehole. The potential applications of surface NDE techniques for pole or pile depth determination are Sonic Echo/Impulse Response (compressional wave generated from longitudinal impulse), Bending Wave (dispersive wave generated from transverse impulse), and Ultraseisminc Vertical Proofing with geophysical processing of the data (compressional and flexural wave). Borehole NDE techniques include Parallel Seismic, i.e., direct measurement of compressional and shear wave arrival to receivers in a borehole emitted by wave travelling down the pile from an impact to the exposed substructure, Borehole Radar, i.e., reflection of electromagnetic wave energy is measured from nearby substructure.

The research has mainly been focused on surface wave methods to meet the time- and cost-effective requirements of the project.

Experimental Investigations

A number of tests have been conducted, including investigations into suitable hardware (such as data acquisition system and cost-effective sensors), stress wave patterns, reliability and uncertainty issues. Three types of columnar specimens have been used in the current investigations, i.e., a 5 m steel RHS beam, a 5 m rectangular timber beam and a 5 m timber pole. Considering the uncertainty on the material property of timber, a steel beam, consisting of homogeneous material without defects, was chosen as benchmark material for all stress wave tests. The rectangular timber beam without any defects was used to verify the behaviour of "ideal" timber during tests. Several timber poles taken out of service and containing local defects and uncertainty on its properties, have been used as "real" pole specimens.

To be able to simulate geological embedment conditions for field specimens, a container, 1.2m x 1.2m in cross-section and three meters in height has been designed and fabricated to contain sand/soil for embedding different type of specimens (see Figure 15(b) and (c)). Fully instrumented benchmark specimens and pole specimens are then embedded, up to three meters, into the foundations consisting of eight layers of sand/soil. Various NDE techniques are then conducted on embedded specimens. The length of embedment can also be changed during the tests by simply pulling the specimens up and compacting sand/soil. Underground or embedded portion of the specimens were also instrumented so that stress wave patterns can be tracked down. Figure 15(a) shows the schematic NDE test set up under embedded conditions.

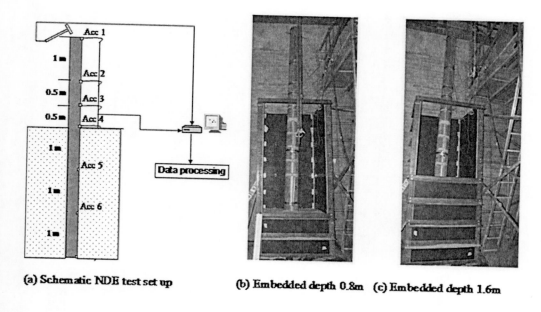

(a) Schematic NDE test set up   (b) Embedded depth 0.8m   (c) Embedded depth 1.6m

Figure 15. The laboratory set-up for NDT test on embedded depth and condition of timber pole

A set of typical results from a longitudinal stress wave test of a timber pole are shown in Figure 16. Results show measured acceleration time histories and calculated velocity after the filtration from six sensors along the length of the pole. Using different techniques, length and condition of the pole can be estimated.

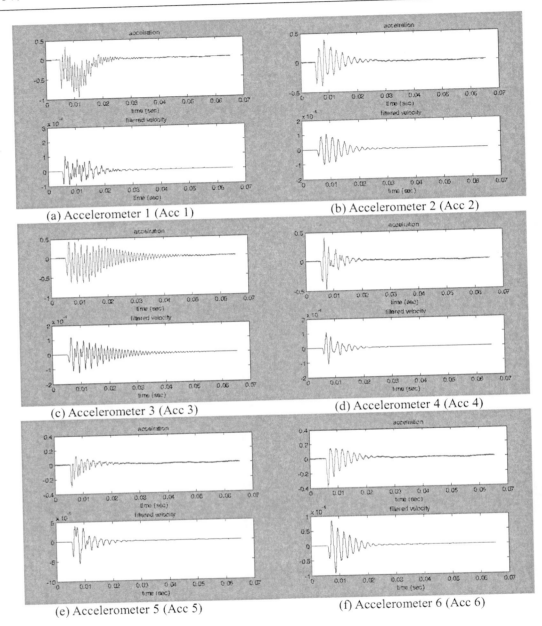

(a) Accelerometer 1 (Acc 1)   (b) Accelerometer 2 (Acc 2)

(c) Accelerometer 3 (Acc 3)   (d) Accelerometer 4 (Acc 4)

(e) Accelerometer 5 (Acc 5)   (f) Accelerometer 6 (Acc 6)

Figure 16. Acceleration and velocity obtained from a timber pole stress wave test

CONCLUSION

This chapter presents research work on vibration-based damage detection methods of timber structures. The developed damage identification procedures are verified by two laboratory structures, a simple timber beam and a laboratory timber bridge deck. A modified damage index (MDI) method was proposed to locate damage in the timber beam structures. The method utilizes the first five mode shapes of the structure and their derivatives obtained from experimental modal analysis. The results showed that the modified algorithm is effective and reliable in locating damage, even though it comes with some false positives. For severity assessment, ANNs are introduced to compensate the shortcoming of the method to quantify severity of damage in the timber beam. The severity estimator of the DI method is used as an input parameter to a neural network ensemble. The individual networks of the ensemble are trained with modally separated severity estimators in order to take advantage of distinguishable patterns of different vibrational modes and thereby to increase efficiency of network training and produce more accurate and robust damage detection results. The outcomes showed that the severities of all damage cases can be successfully determined by the network ensemble. Further, it was shown that the network ensemble produced more accurate predictions than any of the individual networks.

For the laboratory timber bridge, the research proposed the DI-P method for damage location identification. The method uses the first nine modes obtained from experimental modal analysis to locate single- and multiple-damage in the structure. The locations of all severe damage cases in the single- and the two-damage scenarios can be successfully identified. However, light- and medium-damage cases could not be located reliably. To evaluate the severity of damage in the timber bridge, compressed residual FRFs were proposed and used as damage indicator for neural network inputs. Neural networks were then trained with the FRF-based damage fingerprints to predict the severity of accumulative damage cases. The results showed that for all damage cases, even those of light severity, of the laboratory timber bridge, the proposed neural network ensemble was able to identify the level of damage severity reliably.

As an example, the chapter also introduced current research on determination of in-service condition and embedment depth of timber poles using NDE techniques, including some testing set-up and initial results of experimental investigation.

ACKNOWLEDGMENTS

The authors wish to thank the Centre for Built Infrastructure Research (CBIR), the School of Civil and Environmental Engineering, University of Technology Sydney for supporting this work. The authors would like to thank Prof. Keith Crews for contributing his expertise and experiences in many aspects of this study. Finally, the authors wish to render thanks to staff of UTS Structures Laboratory for their assistance in conducting the experimental works.

REFERENCES

Bakhary, N., Hao, H., Deeks. A. J. (2007) "Damage Detection Using Artificial Neural Network with Consideration of Uncertainties." *Engineering Structures.* 29(11) pp. 2806-2815.

Benitez, M.F. and Li, J. (2002) "Static and Dynamic Evaluation of A Timber Bridge (Cattai Ck. Bridge NSW, Australia)." *Proceedings of the 7th World Conference on Timber*

Engineering. (WCTE 2002), 12-15 August, 2002 Shah Alam, Selangor, Malaysia, pp. 385-393.

Choi, F. C., Li, J., Samali, B. and Crews, K. (2007) "Application of Modal-based Damage Detection Method to Locate and Evaluate Damage in Timber Beams." *Journal of Wood Science*. 53(5) pp. 394-400.

Choi, F. C. (2007) *Assessment of the Structural Integrity of Bridges Using Dynamic Approaches*: PhD Thesis: Faculty of Engineering: University of Technology Sydney: Australia.

Choi, F. C., Li, J., Samali, B. and Crews, K. (2008a) "Application of the Modified Damage Index Method to Timber Beams." *Engineering Structures*. 30(4) pp. 1124-1145.

Choi, F. C., Li, J., Samali, B. and Crews, K. (2008b) "An Experimental Study on Damage Detection of Structures using a Timber Beam." *Journal of Mechanical Science and Technology - MOVIC Special Edition*. 21(6) pp. 903-907.

Cornwell, P. J., Doebling, S. W. and Farrar, C. R. (1999) "Application on the Strain Energy Damage Detection Method to Plate-like Structures." *Journal of Sound and Vibration*. 224(2) pp. 359-374.

Crews, K. (2008) "An Overview of the Development of On-site Assessment for Timber Structures in Australia." *Proceedings of the International RILEM conference - On Site Assessment of Concrete, Masonry and Timber Structures*. (SAComaTiS 2008), 1-2 September, 2008 Varenna, Italy, pp. 1113-1124.

Crews, K., Samali, B., Bakoss, S. and Champion, C. (2004) "Overview of Assessing the Load Carrying Capacity of Timber Bridges Using Dynamic Methods." *Proceedings of the 5th Austroads Bridge Conference*, 19-24 May, 2004 Tasmania, Australia, (published on CD-ROM).

Dackermann, U., Li, J., Samali, B., Choi, F. C. and Crews, K. (2008a) "Experimental Verification of a Vibration-Based Damage Identification Method in a Timber Structure Utilizing Neural Network Ensembles." *Proceedings of the International RILEM conference - On Site Assessment of Concrete, Masonry and Timber Structures*. (SAComaTiS 2008), 1-2 September, 2008 Varenna, Italy, pp. 1049-1058.

Dackermann, U., Li, J., Samali, B., Choi, F. C. and Crews, K. (2008b) "Vibration-Based Damage Identification in Civil Engineering Structures utilizing Artificial Neural Networks." *Proceedings of the 12th International Conference on Structural Faults and Repair*. 10-12 June, 2008 Edinburgh, Scotland, (published on CD-ROM).

Dackermann, U., Li, J. and Samali, B. (2009) "Vibration-based Damage Identification in Timber Structures Utilizing the Damage Index Method and Neural Network Ensembles." *Australian Journal of Structural Engineering*. 9(3) pp. 181-194.

DoTaRS (Department of Transport and Regional Services) (2003) *2002-03 Report on the Operation of the Local Government (Financial Assistance) Act 1995*: Canberra: Commonwealth of Australia.

Duwadi, S. R., Ritter, M. A. and Cesa, E. (2000) "Wood in Transportation Program: An Overview." *Fifth International Bridge Engineering Conference*. 3-5 April, 2000, Tampa, Florida, US, pp. 310-315.

Fang, X. and Tang, J. (2005) "Frequency Response Based Damage Detection Using Principal Component Analysis." *Proceedings of the 2005 IEEE International Conference on Information Acquisition*. 27 June-3 July, 2005 Hong Kong and Macau, pp.407-412 .

Francis, L. and Narton, J. (2006) *Australian Timber Pole Resources for Energy Networks* - *A Review*: Report: Department of Primary Industries and Fisheries: Queensland.

Hu, C. and Afzal, M. T. (2006) "A Statistical Algorithm for Comparing Mode Shapes of Vibration Testing Before and After Damage in Timbers." *Journal of Wood Science.* 52(4) pp. 348-352.

Kasal, B. (2008) "In Situ Evaluation of Timber Structures- Education, State-of-the-art and Future Directions." *Proceedings of the International RILEM conference - On Site Assessment of Concrete, Masonry and Timber Structures.* (SAComaTiS 2008), 1-2 September, 2008 Varenna, Italy, pp. 1025-1032.

Kim, J.-T. and Stubbs, N. (1995) "Model-Uncertainty Impact and Damage-Detection Accuracy in Plate Girder." *Journal of Structural Engineering.* 121(10) pp. 1409-1417.

Ko, J. M., Sun, Z. G. and Ni, Y. Q. (2002) "Multi-stage Identification Scheme for Detecting Damage in Cable-stayed Kap Shui Mun Bridge." *Engineering Structures.* 24(7) pp. 857-868.

Lee, J. J. and Yun, C. B. (2006) "Damage Diagnosis of Steel Girder Bridges Using Ambient Vibration Data." *Engineering Structures.* 28(6) pp. 912-925.

Li, J., Choi, F. C., Samali, B. and Crews, K. (2007) "Damage Localization and Severity Evaluation of A Beam-like Timber Structures Based on Modal Strain Energy and Flexibility Approaches." *Journal of Building Appraisal.* 2(4) PP. 323-334.

Li, J., Samali, B. and Crews, K. I. (2004) "Determining Individual Member Stiffness of Bridge Structures Using a Simple Dynamic Procedure." *Acoustics Australia.* 32(1) pp. 9-12.

Li, J., Samali, B., Crews, K., Choi, F. and Shestha, R. (2005) "Theoretical and Experimental Studies on Assessment of Bridges Using Simple Dynamic Procedures." *Proceedings of the 2005 Australian Structural Engineering Conference - Structural Engineering - Preserving and Building into the Future.* (ASEC 2005), 11-14 September, 2005 Newcastle, New South Wales, Australia, (published on CDROM).

Marwala, T. and Hunt, H. E. M. (1999) "Fault Identification Using Finite Element Models and Neural Networks." *Mechanical Systems and Signal Processing.* 13(3) pp. 475-490.

Marwala, T. (2002) "Damage Identification Using Committee of Neural Networks." *Journal of Engineering Mechanics.*126(11) pp. 43-50.

McInnes, K (2005) *Conserving Historic Timber Bridges*: National Trust of Australia: Victoria.

Morison, A., VanKarsen, C. D., Evensen, H. A., Ligon, J. B., Erickson, J. R., Ross, R. J. and Forsman, J. W. (2002) "Timber Bridge Evaluation: A Global Nondestructive Approach Using Impact Generated FRFs." *Proceedings of the 20th International Modal Analysis Conference.* (IMAC-XX 2002), 4-7 February, 2002 Los Angeles, California, US, pp. 1567-1573.

Morison, A., VanKarsen, C. D., Evensen, H. A., Ligon, J. B., Erickson, J. R., Ross, R. J. and Forsman, J. W. (2003) "Dynamic Characteristics of Timber Bridges as A Measure of Structural Integrity." *Proceedings of the 21st International Modal Analysis Conference.* (IMAC-XXI 2003), 3-6 February, 2003 Kissimme, Florida, US, (published on CD-ROM).

Muller, W. (2008) "GPR Inspection of the World's Longest Timber Bridge." *Proceedings of the 12th International Conference on Structural Faults and Repair.* 10-12 June, 2008 Edinburgh, Scotland, (published on CD-ROM).

Nguyen, M., Foliente, G. and Wang, X. (2004) "State of the Practice and Challenges in Non-destructive Evaluation of Utility Poles in Service." *Key Engineering Materials Journal – Advances in Non-destructive Evaluation.* Transportation Technology Publications 270-273 pp. 1521-1528.

Ni, Y. Q., Zhou, X. T. and Ko, J. M. (2006) "Experimental Investigation of Seismic Damage Identification Using PCA-compressed Frequency Response Functions and Neural Networks." *Journal of Sound and Vibration.* 290(1-2) pp. 242-263.

Perrone, M. P. and Cooper, L. N. (1993) "When Networks Disagree: Ensemble Methods for Hybrid Neural Networks." *Artificial Neural Networks for Speech and Vision*: Chapman and Hall: London.

Peterson S. T., McLean, D. I., Symans, M. D., Pollock, D. G., Cofer, W. F., Emerson, R. N. and Fridley, K. J. (2001a) "Application of Dynamic System Identification to Timber Beams I." *Journal of Structural Engineering.* 127(4) pp. 418-425.

Peterson S. T., McLean, D. I., Symans, M. D., Pollock, D. G., Cofer, W. F., Emerson, R. N. and Fridley, K. J. (2001b) "Application of Dynamic System Identification to Timber Beams II." *Journal of Structural Engineering.* 127(4) pp. 426-432.

Ross, R. J., Brashaw, B. K. and Wang, X. (2006) "Structural Condition Assessment of In-service Wood." *Forest Products Journal.* 56(6) pp. 4-8.

Sahin, M. and Shenoi, R. A. (2002) "Quantification and Localization of Damage in Beam-like Structures by Using Artificial Neural Networks with Experimental Validation." *Engineering Structures.* 25(14) pp. 1785-1802.

Samali, B., Bakoss, S. L., Crews, K. I., Li, J. and Champion, C. (2002) "Assessing the Load Carrying Capacity of Timber Bridges Using Dynamic Methods." *IPWEA Queensland Division Annual Conference.* 23 October, 2002 Noosa, Queensland, Australia, (published on CD-ROM).

Samali, B., Bakoss, S. L., Li, J., Saleh, A. and Ariyaratne, W. (2003a) "Assessing the Structural Adequacy of A 3-span Steel-concrete Bridge using Dynamic Methods: A Case Study." *Proceedings of the 10ᵗʰ International Conference: Structural and Faults and Repair.* 1-3 July, 2003. London, UK, (published on CD-ROM).

Samali, B., Crews, K. I., Li, J., Bakoss, S. L. and Champion, C. (2003b) "Assessing the Load Carrying Capacity of Timber Bridges Using Dynamic Methods." *IPWEA Western Australia State Conference.* 7 March, 2003 Perth, Western Australia, Australia, (published on CD-ROM).

Samali, B., Li, J., Choi, F. C. and Crews, K. (2009) "Application of the Damage Index Method for Plate-like Structures to Timber Bridges." *Structural Control and Health Monitoring.* (Published online 8-July-2009 in Wiley Interscience, *DOI: 10.1002/stc.347*).

Wang, X., Wacker, J. P., Morison, A. M., Forsman, J. M., Erickson, J. R. and Ross, R. J. (2005) *Nondestructive Assessment of Single-span Timber Bridges using A Vibration-based Method*: Research Paper FPL-RP-627: Madison: Forest Products Laboratory.

Wu, X., Ghaboussi, J. and Garrett, J. H. Jr. (1992) "Use of Neural Networks in Detection of Structural Damage." *Computers and Structures.* 42(4) pp. 649-659.

Zad, A. (2009) *Doctoral Assessment Report*: Faculty of Engineering and IT: University of Technology Sydney: Australia.

In: Structural Health Monitoring in Australia
Editors: Tommy H.T. Chan and D. P. Thambiratnam

ISBN: 978-1-61728-860-9
©2011 Nova Science Publishers, Inc.

Chapter 6

Civionics—An Emerging Interdisciplinary Research Area in Australia

Brian Uy[*], Xinqun Zhu[±], Yang Xiang[≠], Ju Jia Zou[‡],
Ranjith Liyanapathirana[¥] , Upul Gunawardana[£],
Laurence Pap and Hirantha Perera

Abstract

The Civionics Research Centre at the University of Western Sydney (UWS) is a multi-disciplinary research group with expertise in Civil/Structural Engineering, Electrical Engineering, Telecommunication Engineering, Construction Engineering and Mechanical Engineering. Multi-disciplinary approaches are being developed to advance the current capabilities of structural health monitoring. In this chapter, some recent activities on research, development and implementation of structural health monitoring for civil infrastructure in the centre are briefly introduced. The content includes wireless sensor network development, advanced signal processing and structural condition assessment.

Introduction

Structural Health Monitoring (SHM) is a highly interdisciplinary area of research focused on developing techniques to detect damage in structures such as buildings, bridges, aircraft, offshore structures (Farrar and Worden, 2007). However, most of the SHM research investigations are conducted within mechanical, civil and aeronautical engineering departments, with little involvement of other specialists. The Civionics Research Centre at

[*] SE, UWS, LOCKED BAG 1797, NSW1797 (b.uy@uws.edu.au)
[±] SE, UWS, LOCKED BAG 1797, NSW1797 (xinqun.zhu@uws.edu.au)
[≠] SE, UWS, LOCKED BAG 1797, NSW1797 (y.xiang@uws.edu.au)
[‡] SE, UWS, LOCKED BAG 1797, NSW1797 (j.zou@uws.edu.au)
[¥] SE, UWS, LOCKED BAG 1797, NSW1797 (r.liyanapathirana@uws.edu.au)
[£] SE, UWS, LOCKED BAG 1797, NSW1797 (u.gunawardana@uws.edu.au)

UWS is a multidisciplinary research group with expertise in Civil/Structural Engineering, Computer Engineering, Electrical Engineering, Telecommunication Engineering, Construction Engineering and Mechanical Engineering. Multi-disciplinary approaches are being developed to advance the current capabilities of structural health monitoring. The research area will initially focus on health monitoring of infrastructure, intelligent infrastructure design, intelligent maintenance and repair of infrastructure. In this chapter, some recent activities on research, development and implementation of structural health monitoring for civil infrastructure in the centre are briefly introduced. The content includes wireless sensor network development, advanced signal processing and structural condition assessment.

WIRELESS SENSOR NETWORKS

Civionics was first proposed by Mufti (Mufti et al., 2005; Rivera et al., 2006) to introduce electronics, in particular electro-photonics, to civil structures. The intention was to provide engineers with the knowledge to build "smart" structures that incorporate structural health monitoring (SHM). Over the ensuing years, the Canadian Network of Centres on Intelligent Sensing for Innovative Structures (ISIS Canada Research Network) published a series of design manuals on how Civionics could be implemented. Fibre optic sensors (FOS) and fiber reinforced polymers (FRP) were integrated into innovative structures for SHM. The manuals covered design specifications for installation of FOS, cables, junction boxes, equipment cupboards and also for data collection, management and storage. Civionics area now includes not only electro-photonics but also other technologies and disciplines for SHM, including wireless communications and sensor networks.

Wireless Sensors and Sensor Networks for SHM

Wireless monitoring has emerged in recent years as a promising technology that could greatly impact the field of structural health monitoring and infrastructure asset management. The wireless sensing unit represents the fundamental building block from which wireless structural monitoring systems for civil structures can be constructed. It integrates wireless communications and mobile computing with sensors to deliver a relatively inexpensive sensor platform. Within a wireless structural monitoring system, each wireless sensing unit will be responsible for three tasks (Lynch, 2007): (i) collection of structural response data, (ii) local interrogation of collected measurement data, and (iii) wireless communication of response data or analysis results to a wireless network that comprises other wireless sensing units.

With wireless sensors offering low installation costs and a distributed computing paradigm, many researchers have considered their use within a structural monitoring system. Lynch and Loh (2006) presented a literature review on wireless sensors and sensor networks for structural health monitoring. Many challenges still lie in the way of improving the capabilities of wireless sensors (Lynch & Loh, 2006 and Jamalipour, 2007). They are as follows: a) the power consumption characteristics of current prototype designs are still too high to consider battery power an attractive power source. Power harvesting entails the use of transducers that convert ambient energy sources (e.g., solar power, thermal, wind and vibration) into usable and storable electrical energy. b) The majority of sensor units are passive devices that only record the response of the structure. Wireless sensors with actuation interface will prove to be even more powerful for monitoring structures for damage. c) Wireless sensor networks should be viewed as a decentralized architecture offering parallel processing of measurement data. The distributed data interrogation schemes should be designed explicitly for the parallelism and decentralization offered by a wireless sensor network. Synchronization of signals is particularly important for accurate structural parameter identification and damage detection.

Many of today's wireless technologies, e.g., IEEE 802.11, Bluetooth, Zigbee, operate on the 2.4GHz and 5GHz Industrial, Scientific and Medical (ISM) frequency bands. IEEE 802.11-WLAN/Wi-Fi (Wireless Local Area Networks/Wi-Fi) is a set of low-tier, terrestrial, network technologies for data communication. It is used for wireless sensor networks. Bluetooth (short-range radio frequency signals operating over the 2.4 GHz unlicensed Industrial, Scientific and Medical (ISM) band) networks have also been used for sensory data distribution and processing. In recent years, a new wireless communication protocol has merged that has been explicitly designed for wireless sensor networks, called IEEE 802.15.4. This wireless personal area networks (WPAN) standard provides mobile battery-dependent devices a wireless media access protocol of low complexity (Institute of Electrical and Electronics Engineers, 2003). This wireless standard is intended for use in energy-constrained wireless sensor networks because of its extreme power efficiency.

Research Aim

The aim of this research is to investigate novel techniques for remote monitoring and communication of wireless sensor data and surveillance images over broadband wireless networks. Instead of the conventional way of terminating all sensor cables and remote camera outputs at a single point, we will concentrate on the use of distributed sensor nodes along with passive wireless sensors. Wireless sensor nodes can be placed at a limited number of strategically important points for passive sensor interrogation, data acquisition and processing. Recent terrorist countermeasures and reconstruction of buildings after natural disasters require distributed secure networks of sensors that can be deployed without wires or cabling (Qian et al., 2007).

The amount of research carried out in the structural health monitoring (SHM) area during the past five years shows that this is a problem of significance to both structural and electrical engineers. With the increase in construction of high-rise buildings and long-span bridges, especially in coastal and seismically active areas, SHM becomes important not only for minimizing over-design and the associated wastage of resources but also for real-time monitoring of the performance over the life of the structure. No attempt has been made so far to monitor structures via broadband wireless networks incorporating passive wireless sensors in a low-cost, reliable and robust system. The investigation of passive wireless sensors operating in the ISM band with patch antenna systems is innovative. Also, the design and construction of a proof-of-concept remote monitoring system using broadband wireless networks is novel. Ultimately, the project will lead to a better understanding of the use of wireless sensor networks for real-time structural health monitoring.

The research to be carried out in this project can be divided into two strands. In the first strand, passive, unpowered or very-low power wireless sensors will be investigated for incorporation in structures at the design and construction stage. During the second strand, we will design and build a prototype system to monitor such sensors remotely over the existing 3G wireless network.

Passive Wireless Sensors

The availability of low-power and passive unpowered sensors will make structural health monitoring (SHM) successful in future smart buildings. Also, the sensors will need to be interrogated (in the case of passive sensors) and monitored over secure networks. In this part, we will investigate novel wireless sensors for measuring strain in structural components. While traditional metal foil strain gauges provide high resolution, the wireless sensors to date exhibit low resolution and are susceptible to severe interference from surrounding structures. This is especially problematic in steel structures and reinforced concrete buildings. Passive wireless sensors that can be interrogated through the use of radio frequencies in the unlicensed ISM bands will be investigated theoretically and experimentally. In the literature, passive sensors made of metal cylinders have been modelled by one-port high-Q resonant circuits (Chuang et al., 2005). The sensor is embedded within a concrete structure and interrogated by sending a radio signal of suitable amplitude and phase characteristic. We will investigate the use of pulsed modulation as a means of obtaining a high-resolution output from the resonant cavity of the sensor. Once the resonant frequency has been established to an adequate accuracy, it will be possible to compute the strain by referring to the unstrained (no load) resonant frequency of the cavity resonator.

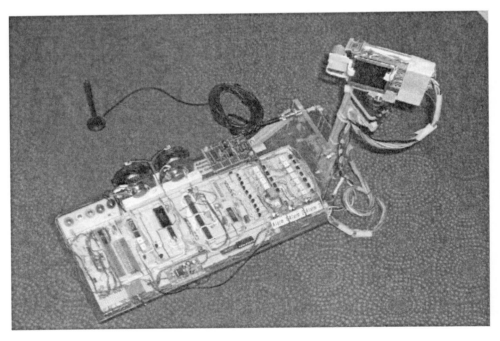

Figure 1. Simple remote monitoring system with built-in videophone

Passive sensors based on RFID (Radio Frequency Identification) tags have also received some attention in the literature (Brocato, 2006; Pohl & Reindl, 1999). In these devices, Surface Acoustic Wave (SAW) technology-based transducers and reflectors are combined with patch antennas for the remote monitoring of large number of sensor-tags. They not only identify the particular sensor but also provide a measure of the attached sensors. In this project, we investigate the feasibility of modifying the conventional RFID tags to respond to strain. These devices are expected to operate under highly noise-prone environments and overcoming channel impairments would be a challenging task for any remote SHM designer.

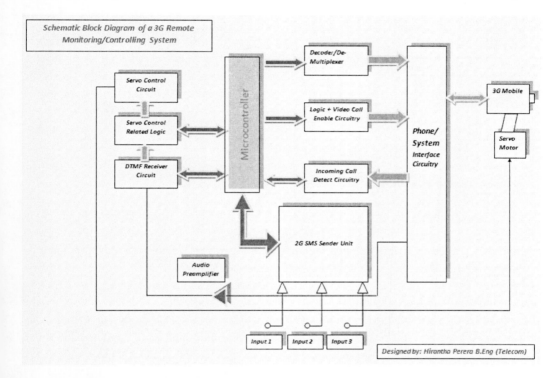

Figure 2. Schematic block diagram of the remote monitoring device

Remote Monitoring and Control Over 3G Wireless Network

The objective of this strand is to design and a build a prototype system that could monitor any number of sensors (multiple types including passive sensors (Chuang et al. 2005) and optical fiber sensors (Lopez-Higuera 2002)) along with surveillance image signals obtained via a small video camera that can be remotely controlled through the 3G telephone network. It will contain built-in intelligence to initiate calls (can be video or voice calls) as a result of predefined adverse events, for example, strain gauge or displacement sensor exceeding the normal operating range or a particular surveillance camera detecting an unauthorized entry to the SHM area. The operator will be able to manipulate the camera over the 3G wireless network for a more detailed view or to focus on an offending structural component. In the normal operating mode, the device will gather data for periodic transmission to a central monitoring station or maintenance centre. Figure 1 shows a simple remote monitoring system built with two 3G mobile phones. The design of the system includes the following tasks: System Controller, Data Acquisition and Interface Design, Image/Video Detection, Image/Video Capturing Software, PCB Design. The system controller consists of a microcontroller to control the various signal processing operations, monitoring functions and video camera remote control operations. The interface will enable the connection of optical fiber sensors and metal foil strain gauges to the system. Sensor data outputs of individual sensors will be sent via a simple SMS whereas images will be sent by initiating a video call to the 3G wireless network. The schematic block diagram of the system is shown in Figure 2. A video, voice or text (SMS) message could be initiated in response to an alarm situation or for routine inspection of the structure. A stepper motor control unit has been designed to operate the camera remotely. Image/Video detection unit along with the video capturing software will enable intruder detection and transmittal of more detailed views of the structure to the operator. All of the above subsystems are combined into compact PCBs within the prototype unit.

ADVANCED SIGNAL PROCESSING

Advanced Image Processing for SHM

Signal processing is an important element of any structural health monitoring system. Various damage detection techniques, approaches and methodologies have been developed over the past 30 years. Many recent studies in this are related to new developments in advanced signal processing (Staszewski and Robertson, 2007). Most damage detection methods use Fourier analysis as the primary signal processing tool. From the resulting spectra, modal properties or damage-sensitive features are extracted to detect change in the signal properties. However, the natural frequencies and mode shapes are not necessarily good indicators of damage, and it needs a large number of measurements to characterize the mode shapes accurately. Moreover, most methods require the preventive knowledge of the dynamic characteristics of the undamaged structure. It is also necessary to remove the effects of environmental factors such as temperature and humidity from the damage detection method. The non-stationary nature of measured signals produced by mechanical faults or structural damage suggests that time-variant procedures can be used to detect such faults as an alternative approach to the classical, Fourier-based methods. Recently, wavelet analysis has become a widely used signal-processing tool in the field of vibration-based damage detection due to its promising features such as singularity detection, good handling of noisy data, being very informative about damage location/time and extent. An extensive literature survey on wavelet based damage detection is available in the work of Kim and Melhem (2004). Hilbert-Huang transform (Huang et al., 1998) is a new time series method to analyze nonlinear and non-stationary data. Empirical mode decomposition is to extract a custom set of basis functions to describe the vibratory response of a system. Hilbert transform is to extract the instantaneous frequency, amplitude and damping. From the time-frequency feature, structural damage can be detected.

Measurement of static and dynamic displacement response is of great importance for assessing the performance, safety and integrity of civil engineering structures. Photogrammetry is a measurement technique that uses photographs to establish the geometrical relationship between a three-dimensional (3D) object and its two-dimensional (2D) photographic images. When sequences of images are used to capture a spatial coordinate time history of the object, the technique can also be referred to as a video-grammetric technique. Fu and Moosa (2002) used monochrome images acquired from a CCD camera to measure static deformation profile of a simply supported beam. A sub-pixel edge detection technique was used together with curve fitting to locate damage in the beam. Jauregui et al. (2003) used a semi-metric digital camera to measure vertical static deflection of bridges. A set of control points with precise 3D coordinates was used to establish the geometrical relationship between targets and their images. Patsias and Staszewski (2002) used a video-grammetric technique to measure mode shapes of a cantilever beam. A wavelet edge detection technique was adopted to detect the presence of damage in the beam. A high-speed professional camera system equipped with a maximum sampling rate of 600 frames per second (fps) was used to capture the beam vibration. The system, however, was limited to 2D planar vibration measurement since only one camera was used. Poudel et al. (2005) used digital video imaging to measure the structural vibration based on sub-pixel edge identification. A laboratory test was carried out on a simply supported beam using a high-speed digital video camera. Chang and Ji (2007) presented a video-grammetric technique to measure three-dimensional structural vibration response in laboratory.

Optical measurement techniques, which are an application of Civionics, have been developed to overcome the complexities of traditional approaches. Optical measurement techniques involve the use of photogrammetric equipment to acquire data from a structure in the form of an image. Image processing can then be applied to the image to overcome the physical limitation of photogrammetric equipment through software. When image processing, such as edge detection, is applied to structural health monitoring, it is important to locate the edges and boundaries of an object as accurately as possible to ensure that correct data is received and correct decisions are made. There have been a number of different edge detection algorithms that have been developed such as Prewitt, Sobel and Canny and are widely accepted and used (Canny, 1986). However, one major problem to overcome is the offset of an actual edge location on the image, which is caused by the quantization of straight lines. These offsets diminish the accuracy of the edge detection algorithms, and therefore incorrect locations of the edges may result. Therefore, to solve this critical problem, sub-pixel edge detection techniques can be preformed on the digital images.

There have been many different techniques developed to estimate the sub-pixel edge positions; these include a method that locates discontinuities in images by using a zero crossing detector (Binford and MacVicar-Whelam, 1981), methods that use spatial or Zernike moments of an edge using edge operators (Overington and Greenway, 1987; Lyvers et al., 1989), a method that uses an analogue-based approach from the first derivative of an output from a charge-coupled device (CCD) image for a fast sub-pixel level edge detection (Ohtani and Baba, 2001), a method that uses a Zernike moment combined with Sobel operators to find a sub-pixel approximation to improve accuracy (Qu et al., 2004) and methods that use image interpolation techniques obtained from a polynomial to estimate a sub-pixel edge approximation (Kubota et al., 2001; Xie et al., 2005). Image interpolation techniques provide an accurate sub-pixel approximation by estimating values that are at undefined points. These techniques are particularly useful in low-resolution images and structural health monitoring, where cost-effective photogrammetric equipment can be purchased to produce similar accuracy to photogrammetric equipment which are expensive.

A popular non-linear method called the Essentially Non-Oscillatory (ENO) interpolation was originally developed to overcome the problem of smoothing across discontinuities in fluid dynamics applications and was later modified for sub-pixel image processing applications (Harten et al., 1987; Balaguer et al., 2001). The ENO interpolation method selects a stencil for a polynomial calculation based on the grey-level intensity variation of a pixel and its surrounding pixels. This is done by computing a standard Newton divided difference of the pixels in the image to form a measure of their smoothness. Two contiguous sets of points are then formed using the divided differences. The contiguous set of points with the lower divided differences is added to stencil. A recent modification of the ENO method was recently developed and uses a fourth-order interpolation ENO method to achieve a finer grid of pixels to increase the geometric accuracy for edge detection (Hermosilla et al., 2008). This method considers the pixels of an image as either point values or cell averages of a function for the purpose of improving the localization and geometry of the edges at a sub-pixel level. A gradient-based edge detector can then be applied to the image to attain the sub-pixel edge location. These methods provide good localization and good detection where edges are marked relatively close to the real edges of an image. For the application of structural health monitoring, it is important to locate the edges of a real object as accurately as possible to ensure that correct data is received and correct decisions are made.

This section presents a new edge detection algorithm at the sub-pixel level for the purpose of structural health monitoring. The proposed method is based upon the ENO interpolation method and follows the scheme introduced by Hermosilla et al. (2008). The proposed method also integrates the Canny edge detector with the ENO interpolation scheme to provide a more accurate and efficient determination of the edges of an object than previous edge detectors.

Edge Detection Algorithm

The proposed method is based upon the ENO interpolation method; however, unlike previous methods, which apply a gradient based edge detector after interpolation, the proposed method combines a fourth-order ENO interpolation method with the Canny edge detector. This is achieved by a series of ENO interpolation calculation on the pixel centre intensities of a cell averaged image and an up-scaled image to determine the local maximums. Using the half pixel centre intensities from the up-scaled image, a determination is made as to which pixel center intensity in the ENO cell averaged image should be considered the real edge. The pixel intensity that is closer to the half pixel centre intensity is then set as the local maximum.

When an object is digitalized, it is represented by a set of pixels that are used to form a two dimensional image matrix. The pixels of an image are represented as the grey level cell average function $\overline{\overline{g}}_{i,j}$ according to Harten (1987) and Hermosilla et al. (2008):

$$\overline{\overline{g}}_{i,j} = \frac{1}{h^2} \int_{y_j - \frac{h}{2}}^{y_j + \frac{h}{2}} \int_{x_i - \frac{h}{2}}^{x_i + \frac{h}{2}} g(x,y)\,dx\,dy \tag{1}$$

where $g(x,y)$ is the discrete grey level value of an image, x and y are the coordinates of a pixel, i and j are the coordinates of a cell, and h is the size of a cell.

The first step of the proposed method is to increase the resolution of the pixels in the image. This is done by dividing a cell average into four new sub-cell averages $\overline{\overline{g}}_{i,j}^{(n)}$ where $n \in \{1,2,3,4\}$. This can be computed according to the procedure described in the paper written by Hermosilla et al. (2008). Figure 3 illustrates the new resolution obtained from dividing the cell averages into four new sub-cell averages.

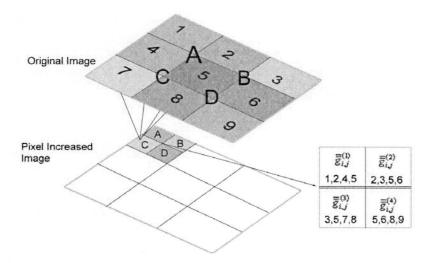

Figure 3. The process of dividing a cell average into four new sub-cell averages

The second step of the proposed method is to remove noise from the new sub-cell averages. This is an addition to the original method by Hermosilla et al. (2008) and is required to increase the performance. The removal of noise before interpolation decreases the likelihood of false readings of the edges of the object, thus improving the accuracy of the method. A mean convolution filter is applied to the cell average values to smooth the edges of the image and to remove noise.

The third step of the proposed method is to perform a sub-pixel estimation using the filtered cell averages $\overline{\overline{G}}_{i,j}$ and a 4th order ENO interpolation. The ENO interpolation stencil $\{\overline{\overline{G}}_{k\min,j}, \overline{\overline{G}}_{k\min,j} +1, \overline{\overline{G}}_{k\min,j} +2, \overline{\overline{G}}_{k\min,j} +3\}$ is formed from four successive points in the x horizontal direction using the filtered cell averages. This stencil is repeated for all the data sets in the x horizontal direction. The data set is also doubled to allow for a higher precision for sub-pixel calculations.

The standard Newton divided differences can be computed according to Harten et al. (1987). The pixel intensity values from the cell average image in either the horizontal or vertical direction make up the first column of the table. Two nearest values are selected from the first column, and the subtraction is stored in column 2. This process is repeated until four columns are computed. The entire table is then used as a measure of the smoothness of intensity values in the image.

Following the method of a standard ENO interpolation, the possible points with the lower divided difference are then added to the stencil. An ENO polynomial according to Hermosilla et al. (2008) is evaluated to form a sub-pixel approximation.

Using the set of sub-pixel approximated values calculated for the x horizontal direction, the ENO procedure is then repeated for values in the y vertical direction. The ENO interpolation stencil $\left\{\overline{\overline{G}}_{j_{min,x_*}}, \overline{\overline{G}}_{j_{min,x_*}} + 1, \overline{\overline{G}}_{j_{min,x_*}} + 2, \overline{\overline{G}}_{j_{min,x_*}} + 3\right\}$ is formed from four successive points in the y vertical direction.

The fourth step of the proposed method is to perform a modification of the Canny Edge detector (Canny, 1986). The calculated sub-pixel approximations are used during the non-maximum suppression stage of the Canny edge detector to suppress non-maximum values, which are not considered to be an edge, to zero. The proposed method differs from the Canny edge detector and Hermosilla et al. (2008) in that the local maximum that is used to suppress non-maximum values is found by using the up-scaled image and determining which pixel intensity values in the ENO average image are closer to the half pixel intensity values. The advantage of using the up-scaled image is that a greater resolution and precision are available in determining which pixels should be considered a real edge.

Finally, $O\left(x_*, y_*; \overline{\overline{G}}\right)$ denotes the output image, which gives the coordinates of the edge points at the sub-pixel level.

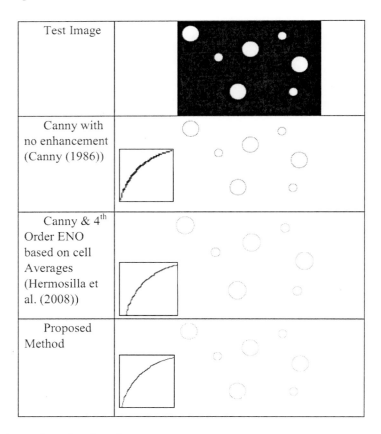

Figure 4. Experimental results of test image.

Experimental Results

The proposed method was tested using various synthetic images and is evaluated against previous method to show its superiority. In image processing, an ideal reference image is non-existent due to the discretization of real values when an object is digitalized. However, synthetic images with known geometry shapes can be used to compare the accuracy of edge detection methods to the theoretical calculated edges of a known geometry shape. Figure 4 illustrates the experimental results of the proposed method when compared to the Canny edge detector with no sub-pixel estimation and with a fourth-order interpolation based on cell averages (Canny, 1986; Hermosilla et al., 2008). It can be seen that the detected edges in the proposed method are better defined and less staggered than the previous two methods. Table 1 illustrates the calculated root mean squared error of the estimated edge locations from each method using the circle synthetic image. It can be seen that the fourth-order ENO method with Canny edge detection provides a vast improvement over Canny without any interpolation. This is due to the sub-pixel estimations in the interpolation process. It can also be seen that the best results are achieved by the proposed method. The computational time for each method is also recorded. The experimental results were preformed on an Intel Core 2 Quad CPU at 2.4GHz with 2 GB or ram and using the C# programming environment. It can be seen that both the proposed method and the fourth-order ENO method are slower than Canny's method. However, they provide a more accurate result. The proposed method is slightly faster than the fourth-order ENO method, and this due to its integration into Canny. Table 2 shows the coordinates of the circle centers for the test image given in Figure 4, calculated by different methods. Again, it can be seen that the best results are achieved by the proposed method.

Table 1. Analytical results for a circle synthetic image

Method	Accuracy (Root mean squared error)	Computational Time (Seconds)
Canny (Canny, 1986)	1.59	0.016
4th Order ENO Canny (Hermosilla et al., 2008)	0.60	2.203
Proposed Method	0.48	2.140

Table 2. Circle centers for test image

	Proposed Method	Canny (Canny, 1986)	4th Order ENO Canny (Hermosilla et al., 2008)	Real Edges (Synthetic)
#1	80.75, 79.75	78, 78	79.75, 80.75	80, 80
#2	450.75, 183.75	448, 182	450.75, 184.75	451, 184
#3	756.75, 283.75	750, 282	753.25, 284.25	756, 284
#4	373.75, 458.75	371, 457	372.75, 459.25	374, 457
#5	252.25, 237.25	250, 236	253.25, 238.25	252, 238
#6	645.25, 96.25	643,95	644.25, 953.75	645, 96
#7	713.25, 461.75	711, 460	712.75, 462.25	713, 462

CONDITION ASSESSMENT

Condition Assessment in Operating Conditions

One of the questions that is attracting significant research attention is related to the use of structural response from operational dynamic loads in a damage detection procedure. For bridges, the operational loads are moving vehicular loads, and the operational deflection shapes are the deflections of the bridge deck subject to moving vehicular loads. Mazurek and Dewolf (1990) conducted the laboratory studies on simple two-span girders under moving loads with structural deterioration by vibration analysis. Structural damages were artificially introduced by a release of supports and insertion of cracks. Majumder and Manohar (2003) developed a time domain formulation to detect damages in a beam using data originating from the linear beam-oscillator dynamic interaction and extended the capabilities of this formulation to include the possibility of the damaged beam structure undergoing nonlinear vibrations. The study combines finite element modeling for the vehicle-bridge system with a time-domain formulation to detect changes in structural parameters. The structural properties and motion characteristics of the moving vehicle are assumed to be available, and the elemental stiffness loss is used to simulate the different damage scenarios. Sieniawska et al. (2009) presented a method to identify the flexural stiffness of a linear structure from displacement measurements using a moving load. Recently, Zhu and his collaborators (Zhu and Law, 2006; Zhu and Law, 2007 and Law and Zhu, 2009) developed algorithms for condition assessment of highway bridges in operating conditions. The following section is a brief summary of the algorithm.

Vehicle-Bridge Interaction Analysis

Vehicular Load Model

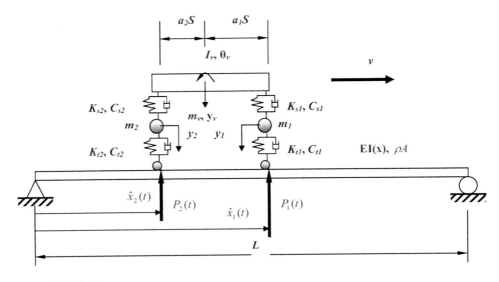

Figure 5. Vehicle-bridge system

A four degrees-of-freedom (DOFs) vehicle model is presented in Figure 5. The equations of motion of the vehicle are derived using Lagrange formulation as follows (Law and Zhu, 2004)

$$\mathbf{M}_v\ddot{\mathbf{Y}} + \mathbf{C}_v\dot{\mathbf{Y}} + \mathbf{K}_v\mathbf{Y} = \mathbf{P(t)} + \mathbf{P_0}$$

(2)

where $\mathbf{Y}$ is the response vector of the vehicle; $\mathbf{P(t)}$ is the vehicle-bridge interaction force vector, and $\mathbf{P_0}$ is the static vehicular load acting on the bridge. $\mathbf{M}_v, \mathbf{C}_v$ and $\mathbf{K}_v$ are the mass, damping and stiffness matrices of the vehicle, respectively.

Bridge Model

The bridge-vehicle system is modelled as a continuous uniform beam subject to a system of moving forces $P_l(t)$ $(l=1,2,...N_p)$, as shown in Figure 5. The forces are assumed moving as a group at a prescribed velocity $v(t)$, along the axial direction of the beam from left to right. The beam is assumed to be an Euler-Bernoulli beam with simple supports. The equation of motion can be written as

$$\rho A\frac{\partial^2 w(x,t)}{\partial t^2} + C\frac{\partial w(x,t)}{\partial t} + \frac{\partial^2}{\partial x^2}\left[EI(x)\frac{\partial^2 w(x,t)}{\partial x^2}\right] = \sum_{l=1}^{N_p} P_l(t)\delta(x - \hat{x}_l(t))$$

(3)

where L is the total length of the beam; A is the cross-sectional area; ρ, C and $w(x,t)$ are the mass per unit length, the damping and the displacement function of the beam, respectively; $\hat{x}_l(t)$ is the location of moving force $P_l(t)$ at time t; $\delta(t)$ is the Dirac delta function and N_p is the number of forces; $EI(x)$ is the flexural rigidity of the beam. Express the transverse displacement $w(x,t)$ in modal coordinates

$$w(x,t) = \sum_{i=1}^{n}\phi_i(x)q_i(t)$$

(4)

where $\phi_i(x)$ is the mode shape function of the ith mode, which is determined from the eigen value and eigen function analysis; $q_i(t)$ is the ith modal amplitude. Substituting Eqn. 4 into Eqn. 3, and multiplying by $\phi_i(x)$, integrating with respect to x between 0 and L, and applying the orthogonality conditions, we obtain

$$\mathbf{M}_b\ddot{\mathbf{Q}} + \mathbf{C}_b\dot{\mathbf{Q}} + \mathbf{K}_b\mathbf{Q} = \mathbf{\Phi}\mathbf{P(t)}$$

(5)

where

$$\boldsymbol{\Phi} = \begin{bmatrix} \phi_1(\hat{x}_1(t))/M_1 & \phi_1(\hat{x}_2(t))/M_1 & \cdots & \phi_1(\hat{x}_{Np}(t))/M_1 \\ \phi_2(\hat{x}_1(t))/M_2 & \phi_2(\hat{x}_2(t))/M_2 & \cdots & \phi_2(\hat{x}_{Np}(t))/M_2 \\ \vdots & \vdots & \vdots & \vdots \\ \phi_N(\hat{x}_1(t))/M_n & \phi_N(\hat{x}_2(t))/M_n & \cdots & \phi_N(\hat{x}_{Np}(t))/M_n \end{bmatrix} \tag{6}$$

$\mathbf{M}_b, \mathbf{C}_b, \mathbf{K}_b$ are mass, damping and stiffness matrices of the beam structure. n is the number of the modes. ω_i, ξ_i, M_i are, respectively, the modal frequency, the damping ratio and the modal mass of the ith mode, and

$$M_i = \int_0^L \rho A \phi_i^2(x)dx \tag{7}$$

Bridge-Vehicle Interaction Forces

The interaction forces between the vehicle and the bridge are given by

$$\begin{cases} P_1(t) = k_{t1}(y_1 - z_1) + c_{t1}(\dot{y}_1 - \dot{z}_1) + (m_1 + a_2 m_v)g \\ P_2(t) = k_{t2}(y_2 - z_2) + c_{t2}(\dot{y}_2 - \dot{z}_2) + (m_2 + a_1 m_v)g \end{cases} \tag{8}$$

where $k_{t1}, k_{t2}, c_{t1}, c_{t2}$ are the stiffness and the damping of the two tires. The displacements underneath the tires z_1 and z_2 are given by

$$\begin{cases} z_1 = w(\hat{x}_1(t),t) + r(\hat{x}_1(t)) \\ z_2 = w(\hat{x}_2(t),t) + r(\hat{x}_2(t)) \end{cases} \tag{9}$$

$$\begin{cases} \dot{z}_1 = \dot{w}(\hat{x}_1(t),t) + w'(\hat{x}_1(t),t)\dot{\hat{x}}_1(t) + r'(\hat{x}_1(t))\dot{\hat{x}}_1(t) \\ \dot{z}_2 = \dot{w}(\hat{x}_2(t),t) + w'(\hat{x}_2(t),t)\dot{\hat{x}}_2(t) + r'(\hat{x}_2(t))\dot{\hat{x}}_2(t) \end{cases} \tag{10}$$

$$\dot{w}(\hat{x}_l(t),t) = \sum_{i=1}^N \phi_i(\hat{x}_l(t))\dot{q}_i(t); \qquad w'(\hat{x}_l(t),t) = \sum_{i=1}^N \frac{\partial \phi_i(x)}{\partial x} q_i(t)\Big|_{x=\hat{x}_l(t)};$$

$$r'(\hat{x}_l(t)) = \frac{dr(x)}{dx}\Big|_{x=\hat{x}_l(t)}; \qquad \dot{\hat{x}}_l(t) = \frac{d\hat{x}_l(t)}{dt}. \qquad (l = 1,2) \tag{11}$$

where $w(x,t)$ is the displacement of the bridge at a distance x from the left support and at time t, and $r(\hat{x}(t))$ is the road surface roughness at the location of the tire.

Substituting Eqn. 9 into Eqns. 2 and 5, the combined equation of the vehicle-bridge system can be obtained:

$$\mathbf{M}(t)\ddot{\mathbf{D}} + \mathbf{C}(t)\dot{\mathbf{D}} + \mathbf{K}(t)\mathbf{D} = \mathbf{F}(t) \tag{12}$$

where $\mathbf{D} = \begin{Bmatrix} \mathbf{Q} \\ \mathbf{Y} \end{Bmatrix}$; $\mathbf{M}(t), \mathbf{C}(t), \mathbf{K}(t)$ are time-varying matrices of the vehicle-bridge system;

$\mathbf{F}(t)$ is the force acting on the system. Eqn. 12 can be solved step-by-step using the central difference method, and the dynamic responses of the bridge under the moving vehicle can be obtained from Eqn. 4.

Road Surface Roughness

In ISO-8606 (1995) specifications, the road surface roughness is often related to the vehicle speed by a formula between the velocity power spectral density (PSD) and the displacement PSD. The general form of the displacement PSD of the road surface roughness is given as

$$S_d(f) = S_d(f_0) \bullet (f / f_0)^{-\alpha} \tag{13}$$

where $f_0 (= 0.1$ cycles/m$)$ is the reference spatial frequency; α is an exponent of the PSD, and f is the spatial frequency (cycles/m). Eqn. 13 gives an estimate on the degree of roughness of the road by the $S_d(f_0)$ value. This classification is made by assuming a constant vehicle velocity PSD and taking α equals to 2. The ISO specification also gives the power spectral densities for different classes of roads.

Based on this ISO specification, the road surface roughness in the time domain can be simulated by applying the Inverse Fast Fourier Transformation on $S_d(f)$ as follows (Henchi et al., 1998)

$$r(x) = \sum_{i=1}^{N} \sqrt{4S(f_i)\Delta f} \, \cos(2\pi f_i x + \theta_i) \tag{14}$$

where $f_i = i\Delta f$ is the spatial frequency; $\Delta f = \dfrac{1}{N\Delta}$; Δ is the distance interval between successive ordinates of the surface profile; N is the number of data points, and θ_i is a set of independent random phase angle uniformly distributed between 0 and 2π.

Cracked Bridge Beam Models

Open Crack Zone

Abdel Wahab et al. (1999) simulated the damage in a reinforced concrete beam as a reduction in the Young's modulus of material, and the following function was proposed:

$$EI(x) = E_0 I(1 - \alpha \cos^2(\frac{\pi}{2} \frac{|x - l_c|}{\beta L / 2})^m)) \quad (l_c - \beta L / 2 < x < l_c + \beta L / 2) \tag{15}$$

where α, β, m are the damage parameters. l_c denotes the mid-point of the damage zone from the left support. Parameter β characterizes the length of the damaged zone, and it lies in the range between 0.0 and 1.0. Parameter α characterizes the magnitude of the damage, and is between 0.0 and 1.0. If α equals 0.0, the beam is intact. When α equals 1.0, the bending stiffness will vanish at the mid-point of the damage zone in the beam. Parameter m characterizes the variation of the Young's modulus from the centre to the two ends of the damage zone. E_0 is the modulus of the intact beam.

The stiffness of an element with an open crack zone then becomes:

$$k_{bij} = E_0 I [\int_0^{l_c - \beta L/2} \phi_i''(x)\phi_j''(x)dx + \int_{l_c - \beta L/2}^{l_c + \beta L/2} (1 - \alpha \cos^2(\frac{\pi}{2}(\frac{|x - l_c|}{\beta L/2})^m)) \phi_i''(x)\phi_j''(x)dx + \int_{l_c + \beta L/2}^{l} \phi_i''(x)\phi_j''(x)dx_i]/M_i$$

$$(i, j = 1, 2, \cdots, n) \tag{16}$$

Breathing Crack Zone

The mechanism of the stiffness degradation is complicated because of the opening/closure of the micro-cracks, especially under cyclic loading. The modeling of the crack opening/closure behavior can be implemented by the elastic-stiffness recovery during the elastic unloading process from tensile state to compressive state (Lee and Fenves, 1998). Here, an additional parameter, s, is used with the open crack model to describe the degradation process, and Eqn. 15 is written as:

$$EI(x) = E_0 I (1 - s\alpha \cos^2(\frac{\pi}{2}(\frac{|x - l_c|}{\beta L/2})^m)) \quad (l_c - \beta L/2 < x < l_c + \beta L/2) \tag{17}$$

where $0 \le s \le 1$. Crack closure is usually thought of as a bilinear mechanism, where the crack is assumed to be either open ($s=1$) or closed ($s=0$), leading to a step change in stiffness. In reinforced concrete, this mechanism would be very complicated, and an experimental test is used for the study.

Multiple Crack Damage

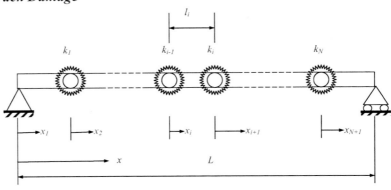

Figure 6. Beam with rotational springs representing damaged section

Figure 6 shows a uniform bridge beam structure with N cracks. The damaged continuous beam is discretized into $N+1$ segments of constant linear density ρA, bending stiffness EI (undamaged beam stiffness) and length l_i, $(i = 1,2,\cdots,N+1)$. The segments are connected together through rotational springs (damage section), whose stiffnesses are denoted by k_i, $(i = 1,2,\cdots,N)$.

The eigen-function of an Euler-Bernoulli beam segment can be written as

$$r_i(x_i) = A_i \sin \beta x_i + B_i \cos \beta x_i + C_i \sinh \beta x_i + D_i \cosh \beta x_i, \qquad (i = 1,2,\cdots,N+1) \quad (18)$$

where $r_i(x_i)$ is the eigen-function for the ith segment, and β is the eigen value of the beam. Zhu and Law (2003) have presented the formulation of eigen-function for a multi-span continuous beam. We have a similar problem with $N+1$ segments connected by rotational springs here. The boundary conditions for the damaged beam are listed as follows:

$$r_1(x_1)\Big|_{x_1=0} = r_{N+1}(x_{N+1})\Big|_{x_{N+1}=l_{N+1}} = 0$$

$$\frac{\partial^2 r_1(x_1)}{\partial x_1^2}\Big|_{x_1=0} = \frac{\partial^2 r_{N+1}(x_{N+1})}{\partial x_{N+1}^2}\Big|_{x_{N+1}=l_{N+1}} = 0$$

$$\left\{ \begin{aligned}
& r_i(x_i)\Big|_{x_i=l_i} = r_{i+1}(x_{i+1})\Big|_{x_{i+1}=0} \\
& \frac{\partial r_i(x_i)}{\partial x_i}\Big|_{x_i=l_i} + \frac{EI}{k_i}\frac{\partial^2 r_i(x_i)}{\partial x_i^2}\Big|_{x_i=l_i} = \frac{\partial r_{i+1}(x_{i+1})}{\partial x_{i+1}}\Big|_{x_{i+1}=0} \\
& \frac{\partial^2 r_i(x_i)}{\partial x_i^2}\Big|_{x_i=l_i} = \frac{\partial^2 r_{i+1}(x_{i+1})}{\partial x_{i+1}^2}\Big|_{x_{i+1}=0} \\
& \frac{\partial^3 r_i(x_i)}{\partial x_i^3}\Big|_{x_i=l_i} = \frac{\partial^3 r_{i+1}(x_{i+1})}{\partial x_{i+1}^3}\Big|_{x_{i+1}=0}
\end{aligned} \right. , \qquad (i = 1,2,\cdots,N) \quad (19)$$

Substituting Eqn. 18 into the boundary conditions Eqn. 19, the mode shape of the continuous beam with N damage locations can be written as

$$\phi(x) = r_1(x)(1 - H(x - l_1)) + \sum_{i=2}^{N+1} r_i(x - \sum_{j=1}^{i-1} l_j)(H(x - \sum_{j=1}^{i-1} l_j) - H(x - \sum_{j=1}^{i} l_j)) \quad (20)$$

where $H(x)$ is the unit step function.

$$\left\{ \begin{aligned}
& r_1(x) = A_1 \sin \beta x + C_1 \sinh \beta x, && (0 \le x < l_1) \\
& r_i(x) = A_i(x) \sin \beta x + B_i \cos \beta x + C_i \sinh \beta x + D_i \cosh \beta x, && (0 \le x < l_i, i = 2,3,...,N+1)
\end{aligned} \right.$$

$$(21)$$

where parameters $\{A\} = \{\beta, A_1, C_1, A_i, B_i, C_i, D_i (i = 2,3,\cdots, N+1)\}$ are determined from the following equation

$$[S]\{A\} = 0 \tag{22}$$

where the elements of matrix S are given by

$$f_{11} = \sin \beta l_1, \qquad f_{12} = \sinh \beta l_1, \qquad f_{14} = -1, f_{16} = -1$$

$$f_{21} = -\sin \beta l_1, \qquad f_{22} = \sinh \beta l_1, \qquad f_{24} = 1, f_{26} = -1$$

$$f_{31} = -\cos \beta l_1, \qquad f_{32} = \cosh \beta l_1, f_{33} = 1, \qquad f_{35} = -1$$

$$f_{41} = \beta \cos \beta l_1 - \beta^2 \frac{EI}{k_1} \sin \beta l_1, \qquad f_{42} = \beta \cosh \beta l_1 + \beta^2 \frac{EI}{k_1} \sinh \beta l_1, \qquad f_{43} = -\beta, f_{45} = -\beta$$

$$f_{4(i-1)-1,4(i-1)-1} = \sin \beta l_i, \quad f_{4(i-1)-1,4(i-1)} = \cos \beta l_i, \quad f_{4(i-1)-1,4(i-1)+1} = \sinh \beta l_i, \quad f_{4(i-1)-1,4(i-1)+2} = \cosh \beta l_i,$$

$$f_{4(i-1)-1,4(i-1)+4} = -1, \qquad f_{4(i-1)-1,4(i-1)+6} = -1, \qquad (i = 2,3,\cdots, N)$$

$$f_{4(i-1)+2,4(i-1)-1} = -\sin \beta l_i, \quad f_{4(i-1)+2,4(i-1)} = -\cos \beta l_i, \quad f_{4(i-1)+2,4(i-1)+1} = \sinh \beta l_i, \quad f_{4(i-1)+2,4(i-1)+2} = \cosh \beta l_i,$$

$$f_{4(i-1)+2,4(i-1)+4} = 1, \qquad f_{4(i-1)+2,4(i-1)+6} = -1, \qquad (i = 2,3,\cdots, N)$$

$$f_{4(i-1)+3,4(i-1)-1} = -\cos \beta l_i, \quad f_{4(i-1)+3,4(i-1)} = \sin \beta l_i, \quad f_{4(i-1)+3,4(i-1)+1} = \cosh \beta l_i, \quad f_{4(i-1)+3,4(i-1)+2} = \sinh \beta l_i,$$

$$f_{4(i-1)+3,4(i-1)+3} = 1, \qquad f_{4(i-1)+3,4(i-1)+5} = -1, \qquad (i = 2,3,\cdots, N)$$

$$f_{4(i-1)+4,4(i-1)-1} = \beta \cos \beta l_i - \beta^2 \frac{EI}{k_i} \sin \beta l_i, \qquad f_{4(i-1)+4,4(i-1)} = -\beta \sin \beta l_i - \beta^2 \frac{EI}{k_i} \cos \beta l_i,$$

$$f_{4(i-1)+4,4(i-1)+1} = \beta \cosh \beta l_i + \beta^2 \frac{EI}{k_i} \sinh \beta l_i, \qquad f_{4(i-1)+4,4(i-1)+2} = \beta \sinh \beta l_i + \beta^2 \frac{EI}{k_i} \cosh \beta l_i,$$

$$f_{4(i-1)+4,4(i-1)+3} = -\beta, f_{4(i-1)+4,4(i-1)+5} = -\beta, \qquad (i = 2,3,\cdots, N)$$

$$f_{4N+1,4N-1} = \sin \beta l_{N+1}, \quad f_{4N+1,4N} = \cos \beta l_{N+1}, \quad f_{4N+1,4N+1} = \sinh \beta l_{N+1}, \quad f_{4N+1,4N+2} = \cosh \beta l_{N+1},$$

$$f_{4N+2,4N-1} = -\sin \beta l_{N+1}, \quad f_{4N+2,4N} = -\cos \beta l_{N+1}, \quad f_{4N+2,4N+1} = \sinh \beta l_{N+1}, \quad f_{4N+2,4N+2} = \cosh \beta l_{N+1}.$$

Crack Identification Using Wavelet Transform

Eqn. 20 shows that there are discontinuities at the damage points, especially the slope discontinuities at the cracks. The damaged locations can be determined by finding the discontinuous points in the mode shapes. Mode shape curvature is widely used to find these discontinuous points (Pandey et al., 1991). However, the first problem for damage detection using curvature directly is to calculate the curvature by derivation. It is very difficult to obtain accurate mode shape in practice, and the differentiation of the mode shape will further amplify the measurement error. Recently, the wavelet transform has been widely used to measure the local regularity of a signal.

The Continuous Wavelet Transform of Measured Displacement

The continuous wavelet transform of a square-integrable signal $f(x)$, where x is time or space, is defined as (Mallat and Hwang, 1992)

$$Wf(u,s) = f(x) \otimes \psi_s(x) = \frac{1}{\sqrt{s}} \int_{-\infty}^{+\infty} f(x)\psi^* \left(\frac{x-u}{s} \right) dx \qquad (23)$$

where $\otimes$ denotes the convolution of two functions. $\psi_s(x)$ is the dilation of $\psi(x)$ by the scale factor s. u is the translation indicating the locality. $\psi^*(x)$ is the complex conjugate of $\psi(x)$, which is a mother wavelet satisfying the following admissibility condition:

$$\int_{-\infty}^{\infty} \frac{|\Psi(\omega)|^2}{|\omega|} d\omega < +\infty \qquad (24)$$

where $\Psi(\omega)$ is the Fourier transform of $\psi(x)$. The existence of the integral in Eqn. 24 requires that

$$\Psi(0) = 0 \text{ i.e., } \int_{-\infty}^{+\infty} \psi(x)dx = 0 \qquad (25)$$

From Eqn. 11, the displacement at x_m can be written as

$$w(x_m,t) = \sum_{i=1}^{\infty} \frac{\phi_i(x_m)}{M_i} \int_0^t h_i(t-\tau)P(\tau)\phi_i(\hat{x}(\tau))d\tau \qquad (26)$$

The second derivation of the displacement with respect to the position of the moving load can be obtained as

$$\frac{\partial^2 w(x_m,t)}{\partial^2 \hat{x}_l(t)} = \sum_{i=1}^{\infty} \frac{\phi_i(x_m)}{M_i} \int_0^t h_i(t-\tau)P(\tau)\frac{\partial^2 \phi_i(\hat{x}(\tau))}{\partial \hat{x}(\tau)^2} d\tau \qquad (27)$$

where $\dfrac{\partial^2 \phi_i(x)}{\partial x^2}$ is the second order derivation of the ith mode, which is the curvature of the displacement mode shape. This shows that the second derivative of the displacement with respect to the load position includes the curvature information of the mode.

Multiscale Differential Operator of Wavelet

Take the Gaussian function $\theta(x)$, and the wavelet of $\theta(x)$ can be defined as the second derivative of the function (Mallat and Hwang, 1992):

$$\psi(x) = \frac{d^2\theta(x)}{dx^2} \qquad (28)$$

The wavelet $\psi(x)$ in Eqn. 28 is continuous differentiable and is usually referred to as the Mexican Hat wavelet that has the following explicit expression:

$$\psi(x) = \frac{2}{\sqrt{3\sigma}} \pi^{-1/4} \left(\frac{x^2}{\sigma^2} - 1 \right) exp\left(\frac{-x^2}{2\sigma^2} \right) \qquad (29)$$

where σ is the standard deviation.

The wavelet transform for the displacement $w(x_m, t)$ is then expressed by the following relation (Mallat and Hwang, 1992) when the Mexican Hat wavelet is adopted, as

$$Ww(\hat{x}(t), s) = w(x_m, t) \otimes \psi_s(\hat{x}(t)) = s^2 \frac{d^2}{d\hat{x}(t)^2} (w(x_m, t) \otimes \theta_s)(\hat{x}(t)) \qquad (30)$$

Eqn. 30 is the multi-scale differential operator of the second order, and is the relation between the second differentiability of $w(x_m, t)$ and its wavelet transform decay at fine scales. The wavelet transform $Ww(\hat{x}(t), s)$ is proportional to the second derivative of $w(x_m, t)$, smoothed by the Gaussian function $\theta_s(x)$. So the wavelet transform can be used to replace the direct differentiation of the displacement to get the curvature properties. The second differential of operating curvature shapes of a beam is not continuously differentiable at the damage location, while in the present case, the measured location is continuously differentiable. The damage can then be located using the wavelet transform of the operational displacement time history at one point when the beam structure is subject to the action of the moving load. Similar formulation can be obtained for accelerations.

EXPERIMENTAL STUDY

Experimental Setup

A Tee-section reinforced concrete beam of 5m length, with 4.8m simply supported span, was tested in this study. The dimensions of the beam are shown in Figure 7. The beam section is 415mm high with 650mm wide flange and 125mm web thickness. A 60mm diameter hole is left in the beam rib with a parabolic profile for possible post-tensioning of the beam. There are five numbers of 16mm diameter mild steel bars at the bottom of the beam, and six numbers of 6mm diameter steel bars at the top of the beam section. The beam was resting on top of 50mm diameter steel bars, which, in turn, rest on top of solid steel supports connected to the concrete floor. The properties of concrete from material tests are: compressive strength is 38.3MPa; density is 2323.5 kg/m^3; tensile strength is 2.63 MPa; Young's Modulus is 23.9

GPa and Poisson's ratio is 0.19. The properties of steel bar are: Young's Modulus is 181.48 GPa and yield stress is 526.33 MPa.

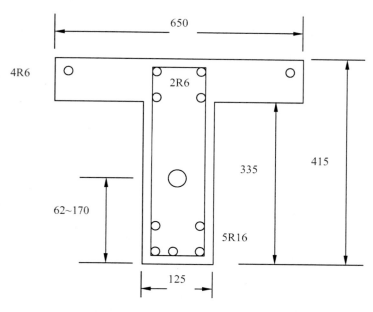

Figure 7. Cross-section layout of the reinforced concrete beam (Dimensions are in mm)

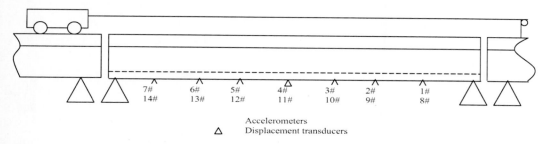

Figure 8. Experimental setup

The experimental set-up shown in Figure 8 includes three Tee-section concrete beams, i.e., the leading beam, the main beam and the tailing beam. The length of the leading and tailing beams are 4.5m each. The gaps between the beams are 10mm. A vehicle is pulled along the beam by an electric motor at an approximate speed of 0.5m/s. The axle spacing of the vehicle is 0.8m, and the wheel spacing is 0.39m. The "light" configuration and the "heavy" configuration of the vehicle weighs 10.60kN and 15.06kN, respectively. For the light vehicle, the front axle load is 5.58kN, and the rear axle load is 5.02kN. For the heavy vehicle, the front axle load and the real axle load are 7.92kN and 7.14kN, respectively. As the total mass of the concrete beam is 1050 kg, the weight ratios between the vehicle and bridge are 1.01 and 1.43 for the two vehicle configurations, respectively.

Seven displacement transducers (sensors 1# to 7#) and seven accelerometers (sensors 8# to 14#) are evenly distributed at the bottom and along the beam to measure the responses as marked in Figure 8. (All the sensors are located at $1/8L$ points along the beam). Thirteen photo-electric sensors are installed on the leading beam and the main beam at 0.56m spacing to monitor the speed of the vehicle. The third and thirteenth photo-electric sensors are located at the entry and exit points of the main beam separately. INV300 data acquisition system is used to collect the data from all channels. The sampling frequency is 2024.292Hz, and the sampling period is 30s for each test.

Test Procedure

The beam is tested when it is undamaged, with a small damage and with a large damage, respectively. The damage is created in the form of cracks with three-point and four-point static load applied to the beam. The following tests are performed for each of the above states of the beam:

- Static load test.
- Vibration response to impact excitation test.
- Moving vehicle load test

Static Load Tests

Damage in the beam is created using a three-point load system applied at *1/3L* from the right support as shown in Figure 9(a). The load is gradually increased at 2 kN increments. When 36kN is reached, several tensile cracks are clearly seen on the beam rib. When the load increases to 50kN, the crack width of the largest crack at the bottom of the beam measured 0.10mm. The location of this crack is close to the loading position but on the inside of the span with a visual crack depth of 213mm and a crack zone of 760mm wide. After the load is kept on the beam for 30 minutes, the beam was unloaded, and the crack closes partly with the crack width at the bottom of the beam reduced to 0.025mm. These observations are for the small damage case. During the loading and unloading procedure, all strain measurements and deflections at quarter span and *2/3L* of the beam are measured.

For the large damage case, the beam is first loaded at *2/3L* of the beam up to 50kN using the three-point load system. This creates a crack pattern similar in magnitude and extent to the existing crack zone at *2/3L*. Further loaded is made using a four-point load system as shown in Figure 9(b). The final total load is 105kN without yielding of the main reinforcement. The largest crack is close to the middle of the beam with 281mm depth. The width of this crack at the bottom of the beam is 0.1mm at 105kN load, and it becomes 0.038mm when the beam is unloaded after keeping the 105kN static load on top for 30 minutes. The crack zone is measured 2371mm long.

Cracks in the concrete beam occur in groups. It would be useful to describe the crack damage in reinforced concrete beam with a crack zone. The damage function proposed by Maeck et al. (2000), which includes three parameters (the central position, the length and the magnitude of damage), would be an effective method to describe the crack zone. In the present case, the small damage is defined as at *2/3L* with 760mm length crack zone and a maximum crack depth of 213mm. The large damage is defined as at *1/2L* with 2371mm length crack zone and a 281mm crack depth.

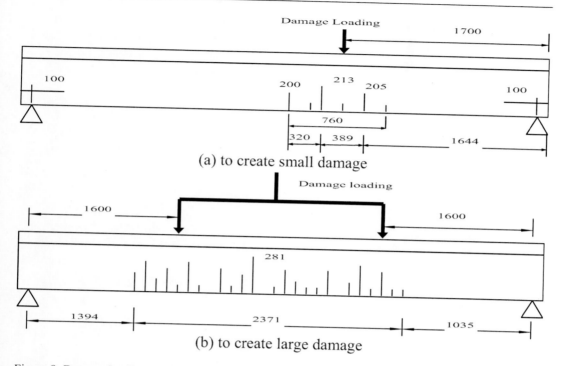

(a) to create small damage

(b) to create large damage

Figure 9. Damage loading and the crack zone (Dimensions are in mm)

Static Vehicular Loads

The 15.06kN vehicle load is put on top of the beam at *1/4L, 1/2L* and *3/4L*, separately. The deflections of the beam are measured by seven displacement transducers distributed evenly under the beam, and they are shown in Table 3. The following observations are made:

Table 3. Beam deflections (mm) under 15.06kN load at different position and the calculated flexural rigidity of beam

Load Position	No damage			Low damage			Damage		
	1/4L	1/2L	3/4L	1/4L	1/2L	3/4L	1/4L	1/2L	3/4L
7/8L	0.128	0.257	0.233	0.245	0.277	0.149	0.284	0.306	0.209
3/4L	0.272	0.480	**0.400**	**0.515**	0.673	0.455	**0.711**	0.851	0.526
5/8L	0.337	0.581	0.474	0.649	0.937	0.684	0.835	1.062	0.668
1/2L	0.475	**0.702**	0.478	0.672	**0.958**	0.758	0.859	**1.137**	0.731
3/8L	0.503	0.668	0.426	0.608	0.978	0.781	0.683	1.060	0.853
1/4L	**0.529**	0.482	0.251	0.471	0.809	**0.705**	0.555	0.889	**0.740**
1/8L	0.258	0.296	0.165	0.210	0.391	0.398	0.276	0.475	0.463
EI ($\times 10^7 \, Nm^2$)	3.672	4.926	4.862	3.696	3.608	2.758	2.733	3.041	2.628

Note: Values **bolded** are used for the calculation of the flexural rigidity of the beam.

1) The deflections become larger, and the stiffness of the beam is reduced when the beam is damaged. The average flexural stiffness can be calculated using the static deflection of the beam at the load position. The damage index is defined in this way:

$$\alpha = 1 - \frac{EI}{EI_{intact}} \tag{31}$$

where EI_{intact} is the flexural stiffness of the undamaged beam. The damage index is 0.268 for the small damage case and 0.361 for the large damage case.

2) Table 3 also shows that the flexural stiffness is different when the vehicle load is located at different positions. When the vehicle load is at $3/4L$ for the small damage case, the calculated flexural stiffness is smaller than that when the vehicle load is located at $1/4L$ or $1/2L$. This is because the vehicle load is close to the location of the crack, causing the crack to open fully. The crack width reduces when the vehicle load is located at other locations along the beam.

Moving Vehicle Loads

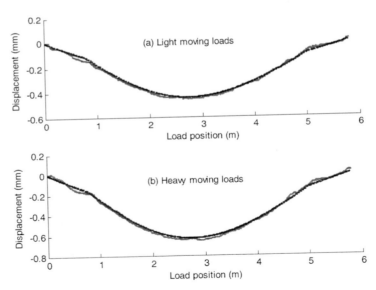

Figure 10. Deflections at $3/8L$ subjected to moving loads (___measured; --- moving loads; ... a moving vehicle)

The damage parameters of the open crack model are taken as: $\alpha = 0.5, \beta = 0.25, l_c = 1/3L, m = 2.0$ for the small damage case and $\alpha = 0.58, \beta = 0.5, m = 2.0, l_c = 1/2L$ for the large damage case. The parameters of the vehicle are: $m_v = 1081.6kg$ and $1536.7kg$ for the light and heavy vehicle, respectively, and I_v, k_s and c_s for each wheel are $2.94 \; kgm^2$, $6.23 \times 10^6 \; N/m$, and $8.94 \times 10^3 \; Nms$, respectively. The road surface roughness is not included in the simulation. The first five modes of the beam are used in the calculation. The bending stiffness (EI) is calculated from the gross cross-section of the beam and equals $5.59 \times 10^7 \; Nm^2$ for the undamaged reinforced concrete section. Figure 10 shows the measured and calculated deflection of the reinforced concrete beam at $3/8L$, subject to the moving vehicle loads without damage in the beam modeled as a set of moving loads or a moving vehicle. Figures 11 and 12 show the measured and calculated deflections of the reinforced concrete beam at $3/8L$, subjected to moving vehicle loads with small and large damages respectively. The following observations are made:

1) The measured and calculated deflections are close to each other. It shows that the damage model is accurate enough to describe the cases of damage in the reinforced concrete beam.

2) The calculated deflections from moving load assumption are approximately the same as those from a moving vehicle assumption. This is due to the low speed of the vehicle, which is about 0.5m/s, and there is very little oscillation when the vehicle crosses the beam.

3) When the damage in the beam increases, the deflection also increases. The deflection would be an effective indicator of damage in the beam.

4) In Figure 11, the measured deflection is larger than the calculated deflection. This indicates a possible under-estimation on the small damage.

5) In Figure 12, the calculated deflection is larger than the measured deflection under the light vehicle. This may be due to the fact that the cracks are not fully opened when the light vehicle is moving on top of the beam.

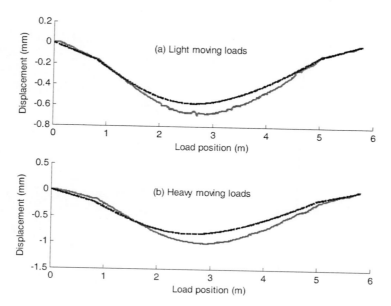

Figure 11. Deflections at $3/8L$ subjected to moving loads with small damage (___measurements; --- moving loads; ... a moving vehicle)

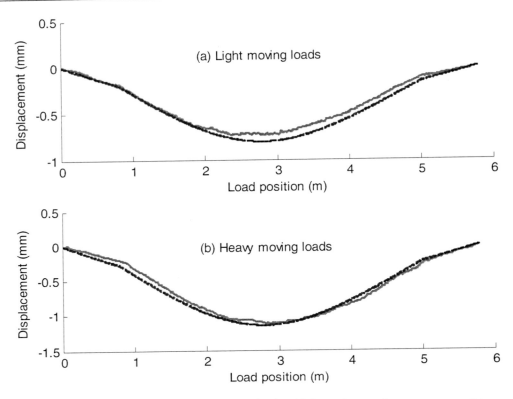

Figure 12. Deflections at *3/8L* subjected to moving loads with large damage (___measurements; ---
moving loads; … a moving vehicle)

Time-Frequency Analysis (Law and Zhu, 2005)

Both types of vehicle are then pulled separately along the beam, and the accelerations and
displacements at *1/8L* to *7/8L* are measured. Figures 10, 11, 12 give the deflection histories at
3/8L for the different states of the beam with the light or heavy vehicle crossing the beam.
The sampling frequency is 2024.292Hz. Figure 13(a) shows the normalized power spectral
density of the acceleration at *1/2L* for the three states of the beam when the light vehicle is at
1/3L. Figure 13(b) shows the normalized power spectrum of the acceleration at *1/2L* when the
light vehicle is at *1/10L, 1/2L* and *9/10L* of the beam with small damage. The frequency
power spectrum is determined using the moving window AR model with an order equal to 21
and 1024 FFT length. Burg (Marple, 1987) algorithm is used in the calculation with the same
length of window as that for the FFT. The peaks for each vibration mode are not very sharp
because of the existence of damping, and the natural frequency is taken either at the location
of the peak or the middle of the frequency range covered by the spectral peak. Figure 14
shows the instantaneous fundamental frequency changes when the vehicle is moving along
the beam. Time-frequency analysis on the collected signal is performed with the windowed
discrete-time Fourier transform algorithm. The FFT length is 1024. The following
observations are made:

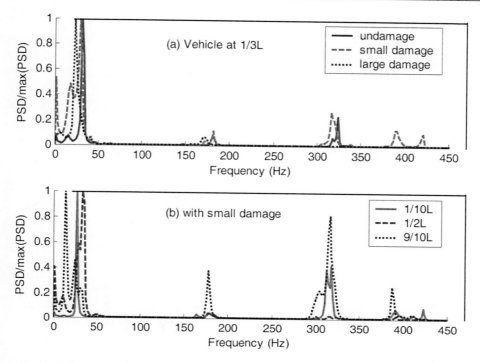

Figure 13. Normalized power spectrum density of the acceleration at *1/2L*

1) The deflection curves in Figure 10 for the undamaged cases are similar to the maximum deflection at mid-span proportional to the mass of the vehicles, which are 10.60 kN and 15.06 kN. This indicates the linearity of the beam under the different loads. In the curves for the small damage case, there are times when the deflection of the beam is relatively constant. The same cracked stiffness of the beam are checked to exist in both cases by adding the linear deflection from the difference of weight of the two vehicles to the deflection of the light vehicle, and that gives the deflection of the heavy vehicle. This phenomenon indicates that the cracks in the beam are fully opened, with the tension in the beam section taken up by the exposed reinforcement in the cracked section.

   Comparison of the curves for the two damage cases with light vehicle moving on top (Figures 11(a) and 12(a)) shows that the "large" damage created from static load does not reappear under the moving load because the two curves have almost the same maximum deflection. The cracks in the large damage case are only partially open for the load moving in the middle length of the beam. There is also a region of maximum deflection in the large damage case with the heavy vehicle, and we cannot determine whether the crack is fully or partially open under the moving vehicular load.

2) Figure 13 shows that the vibration modes are well separated, and there are large shifts in the modal frequencies when the moving vehicle is at different positions, because the cracks open or close when the vehicle is moving along the beam.

3) In Figure 14, the result for the first 0.25 second is not accurate, due to the zero padding at the beginning of the data length for the FFT analysis. The curves show that the net effect with the vehicle moving on the undamaged beam is an increase in

the instantaneous frequency. When the same vehicle crosses the beam with small damage close to the *2/3L*, the stiffness reduction in the beam results in a net decrease in the frequency. When the vehicle passes through the crack zone-I, the beam regains some of the stiffness loss due to the crack opening, and the frequency of the beam goes up again. It is only when the vehicle crosses the beam with large damage that the combined stiffness change in the beam results in a reduction in the instantaneous frequency for most of the time when the vehicle is on the beam.

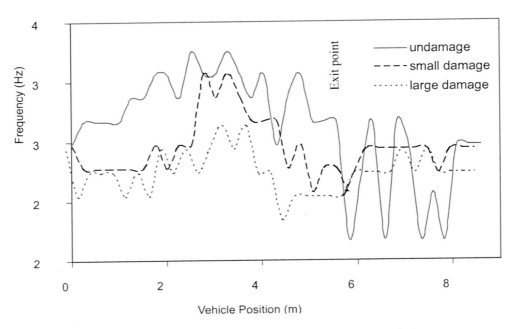

Figure 14. Vehicle position-frequency relation when the vehicle is moving on the beam

Damage Identification by Wavelet Analysis

Figure 15 shows the wavelet coefficients of the displacement at *3/8L* (3# transducer) when the model car was moving on the concrete beam. There are mainly six peaks (X1, X2, X3, X4, X5 and X6 in Figure 15) for the small damage case. The first and second peaks are associated with impacts on the entry of the front and rear axles. The fifth and sixth peaks are associated with impacts on the exit of the front and rear axles. The third and fourth peaks are related to the locations of the damage in the middle figure of Figure 15. The results show that the damage location can be determined using the peaks in the wavelet coefficient of the response from a single measuring point. For the large damage case, there are many cracks created in the reinforced concrete beam. There are also many peaks in the curve of the wavelet coefficient besides those associated with the entry and exit of the vehicle, and the damage zone can be clearly estimated, but the crack location cannot be determined separately. This can be explained by the fact that the large static load of *105* kN has caused bond slippage between the steel bar and concrete, and the damage cannot be simply modelled as an open crack.

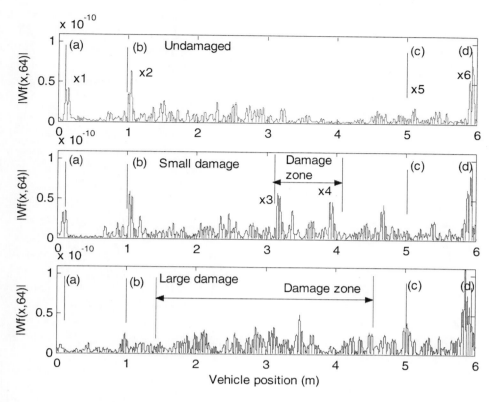

Figure 15. Spatial wavelet coefficient at scale 64 (a) denotes entry of first axle, (b) denotes entry of second axle, (c) denotes exit of first axle, (d) denotes exit of second axle.

CONCLUSION

In this chapter, some recent activities on research, development and implementation of structural health monitoring for civil infrastructure in the UWS Civionics Research Centre are briefly introduced. These activities includes remote monitoring and control over 3G wireless network, development of edge detection algorithms at the sub-pixel level and condition assessment strategy of structures in operating conditions. The next step is to combine these strengths to develop the structural health monitoring system.

REFERENCES

Abdel Wahab M.M., De Roeck G. and Peeters B. (1999). "Parameterization of damage in reinforced concrete structures using model updating." *Journal of Sound and Vibration*, 228(4), pp.717-730.

Balaguer, A, Conde, C, López and Marti´nez, V. (2001). "A finite volume method with a modified ENO scheme using a Hermite interpolation to solve advection diffusion equations." *International Journal on Numerical Methods in Engineering*, 50, pp.2339-2371.

Binford, P.J. and MacVicar-Whelan, P.J. (1981). "Line finding with subpixel precision." *Proceeding of the Society of Photo-Optical Instrumentation Engineers*, 281, pp.26-31.

Brocato R. (2006). "Passive wireless sensor tags." *Sandia Report, SAND2006-1288*, Sandia National Laboratories.

Canny, J. (1986). "A computation approach to edge detection." *IEEE Transaction on Pattern Analysis and Machine Intelligence*, 8, pp.679-698.

Chang C.C. and Ji Y.F. (2007). "Flexible videogrammetric technique for three-dimensional structural vibration measurement." *Journal of Engineering Mechanics ASCE*, 133(6), pp.656-664.

Chuang, J., Thomson, D.J. and Bridges, G.E. (2005). "Embeddable wireless strain sensor based on resonant rf cavities." *Review of Scientific Instruments*, 76, 094703.

Farrar C.R. and Worden K. (2007) "An introduction to structural health monitoring." *Philosophical Transactions of the Royal Society A: Mathematical, Physical & Engineering Sciences*, 365(1851), pp.303-315.

Fu G. and Moosa A.G. (2002). "An optical approach to structural displacement measurement and its application." *Journal of Engineering Mechanics ASCE*, 128(5), pp.511-520.

Harten, A., Engquist, B., Osher, S. and Chakravarthy, S.R. (1987). "Uniformly high order accurate essentially non-oscillatory schemes, III." *Journal of Computational Physics*, 71, pp.231-303.

Harten, A. (1987). "ENO Schemes with Subcell Resolution." *Journal of Computation Physics*, 89, pp.148-184.

Henchi K., Fafard M., Talbot M. and Dhatt G. (1998). "An efficient algorithm for dynamic analysis of bridges under moving vehicles using a coupled modal and physical components approach." *Journal of Sound and Vibration*, 212, pp.663-683.

Hermosilla, T., Bermejo, E., Balaguer, A. and Ruiz, L.A. (2008). "Non-linear fourth-order image interpolation for subpixel edge detection and localization." *Image and Vision Computing*, 26, pp.1240-1248.

Huang N.E., Shen Z., Long S.R., Wu M.C., Shih H.H., Zheng Q., Yen N.-C., Tung C.C. and Liu H.H. (1998). "The empirical mode decomposition and the Hilbert spectrum for nonlinear and non-stationary time series analysis." *Proceedings of the Royal Society of London A: Mathematical, Physical & Engineering Sciences*, 454, pp.903–995.

Institute of Electrical and Electronics Engineers (IEEE), (2003). Part 15.4: Wireless Medium Access Control (MAC) and Physical Layer (PHY) Specifications for Low-Rate Wireless Personal Area Networks (LRWPANs), IEEE, New York, NY.

ISO 8606: 1995(E). Mechanical vibration---road surface profiles---reporting of measured data.

Jamalipour, A. (2007). "Wireless Sensor Applications." *IEEE Communications Magazine*, pp. 2-3.

Jauregui D.V., White K.R., Woodward C.B. and Leitch K.R. (2003). "Noncontact photogrammetric measurement of vertical bridge deflection." *Journal of Bridge Engineering ASCE*, 8(3), pp.212-222.

Kim H. and Melhem H. (2004). "Damage detection of structures by wavelet analysis." *Engineering Structures*, 26, pp. 347-362.

Kubota, T., Huntsberger, T. and Martin, J.T. (2001). "Edge based probabilistic relaxation for sub-pixel contour extraction." *Energy Minimization Methods in Computer Vision and Pattern Recognition, France*, 3, pp.328-348.

Law S.S. and Zhu X.Q. (2004). "Dynamic behavior of damaged concrete bridge structures under moving vehicular loads." *Engineering Structures*, 26(9), pp.1279-1293.

Law S.S. and Zhu X.Q. (2005). "Nonlinear characteristics of damaged concrete bridge structures under vehicular loads." *Journal of Structural Engineering ASCE*, 131(8), pp.1277-1285.

Law S.S. and Zhu X.Q. (2009). *Assessment of Structures under Operating Conditions*. Taylor & Francis, The Netherlands.

Lee J. and Fenves G.L. (1998). "Plastic-damage model for cyclic loading of concrete structures." *Journal of Engineering Mechanics ASCE*, 124(8), pp.892-900.

Lopez-Higuera, J.M. (2002). *Handbook of Optical Fiber Sensing Technology*. Wiley.

Lynch J.P. and Loh K.J. (2006). "A summary review of wireless sensors and sensor networks for structural health monitoring." *The Shock and Vibration Digest*, 38(2), pp.91-129.

Lynch J.P. (2007). "An overview of wireless structural health monitoring for civil structures." *Philosophical Transactions of the Royal Society A: Mathematical, Physical & Engineering Sciences*, 365(1851), pp.345-372.

Lyvers, E.P., Mitchell, O.R., Akey, M.L. and Reeves, A.P. (1989). "Subpixel measurements using a moment-based edge operator." *IEEE Transaction on Pattern Analysis and Machine Intelligence*, 11, pp.1293-1309.

Maeck J., Abdel Wahab M., Peeters B., De Roeck G., De Visscher J., De Wilde W.P., Ndambi J.-M. and Vantomme J. (2000). "Damage identification in reinforced concrete structures by dynamic stiffness determination." *Engineering Structures*, 22(10), pp.1339-1349.

Majumder L. and Manohar C.S. (2003). "A time-domain approach for damage detection in beam structures using vibration data with a moving oscillator as an excitation source." *Journal of Sound and Vibration*, 268(4), pp.699-716.

Mallat S. and Hwang W.L. (1992). "Singularity detection and processing with wavelets." *IEEE Transactions on Information Theory*, 38(2), pp.617-643.

Marple,S.L. (1987). *Digital Spectral Analysis*. Englewood Cliffs, NJ, Prentice-Hall.

Mazurek D.F. and Dewolf J.T. (1990). "Experimental study of bridge monitoring techniques." *Journal of Structural Engineering ASCE*, 115(9), pp.2532-2549.

Mufti A.A., Bakht B., Tadros G. and Horosko A.T. (2005). "Are civil structural engineers 'Risk Averse?' can civionics help?" *Sensing Issues in Civil Structural Health Monitoring*.

Ohtani, K., and Baba, M. (2001). "A Fast Edge Location Measurement with Subpixel Accuracy using a CCD Image." *IEEE Transaction on Instrumentation and Measurement Technology*, 18(3), pp.2087-2092.

Overington, I. and Greenway, P. (1987). "Practical first-difference edge detection with subpixel accuracy." *Image and Vision Computing*, 5, pp.217-224.

Pandey A.K., Biswas M. and Samman M.M. (1991). "Damage detection from changes in curvature mode shapes." *Journal of Sound and Vibration*, 145(2), pp.321-332.

Patsias S. and Staszewski W.J. (2002). "Damage detection using optical measurements and wavelets." *Structural Health Monitoring*, 1(1), pp.7-22.

Pohl A. and Reindl L. (1999). "New passive sensors." *IEEE Xplore* doc no. 0-7803-5276-9/99.

Poudel U.P., Fu G. and Ye J. (2005). "Structural damage detection using digital video imaging technique and wavelet transformation." *Journal of Sound and Vibration*, 286, pp.869-895.

Qian, Y., Lu, K. and Tipper, D. (2007). "A design for secure and survivable wireless sensor networks." *IEEE Wireless Communications,* 14(5), pp.30-37.

Qu, Y.D., Cui, C.S., Chen, S.B. and Li, Q. (2004). "A fast subpixel edge detection method using Sobel-Zernike moments operate." *Image and Vision Computing,* 23, pp.11-17.

Rivera E., Mufti A.A. and Thomson D.J. (2004). "Civionics specifications." *ISIS Canada Design Manual No.6.*

Rivera E., Mufti A.A. and Thomson D.J. (2006). "Civionics specifications for fibre optic sensors for structural health monitoring." *The Arabian Journal of Science and Engineering,* pp.241-248.

Sieniawska R., Sniady P. and Zukowski S. (2009). "Identification of the structure parameters applying a moving load." *Journal of Sound and Vibration,* 319, pp.355-365.

Staszewski W.J. and Robertson A.N. (2007). "Time-frequency and time-scale analysis for structural health monitoring." *Philosophical Transactions of the Royal Society A: Mathematical, Physical & Engineering Sciences,* 365(1851), pp.449-477.

Xie, S.H., Liao, Q. and Qin, S.R. (2005). "Sub-pixel edge detection for precision measurement based on Canny criteria." *Key Engineering Materials,* 295-296, pp.711-716.

Zhu X.Q. and Law S.S. (2003). "Identification of moving interaction forces with incomplete velocity information." *Mechanical Systems and Signal Processing,* 17(6), pp.1349-1366.

Zhu X.Q. and Law S.S. (2006). "Wavelet-based crack identification of bridge beam from operational deflection time history." *International Journal of Solids and Structures,* 43(7-8), pp.2299-2317.

Zhu X.Q. and Law S.S. (2007). "Damage detection in simply supported concrete bridge structure under moving vehicular loads." *Journal of Vibration and Acoustics ASME,* 129, pp.58-65.

In: Structural Health Monitoring in Australia ISBN: 978-1-61728-860-9
Editors: Tommy H.T. Chan and D. P. Thambiratnam ©2011 Nova Science Publishers, Inc.

Chapter 7

STRUCTURAL RELIABILITY ANALYSIS AND SERVICE LIFE PREDICTION FOR STRUCTURAL HEALTH MONITORING

*Mark G. Stewart*and *Hong Hao*[‡]

1 The University of Newcastle, Callaghan, NSW, Australia
2 The University of Western Australia, Crawley, WA, Australia

ABSTRACT

Reliability-based safety assessment is used frequently in Europe and elsewhere to assess the need for maintenance and the remaining service life of structures. In order to assess the effect of corrosion, quantitatively, an experimental study was conducted using an accelerated corrosion testing technique. Vibration tests were carried out fortnightly to study its effect on the natural frequency of RC beams subject to corrosion. One beam was taken out and broken every four weeks. The mass losses of steel rebar were measured to determine the corrosion state. The experimental results are used to develop an empirical model that describes the relationship between natural frequency and corrosion loss. The statistics for model error for this relationship were then inferred. A spatial time-dependent structural reliability analysis was developed to update the deterioration process and evaluate the probability of structural failure based on vibration results. The performance of RC beams is used to illustrate the reliability analysis developed in this paper. The inspection finding considered herein is the timing of test and the vibration results, which are used to provide an updated estimate of structural reliability. It was found that vibration test findings change, the future reliability predictions significantly. A description of structural reliability and reliability-based safety assessment is also provided, where failure probabilities are compared with a typical target failure probability to illustrate how condition assessment findings can be used to more accurately assess and often increase service life predictions.

---

* Centre for Infrastructure Performance and Reliability, School of Engineering, The University of Newcastle, Callaghan, NSW, 2308, Australia (mark.stewart@newcastle.edu.au)
‡ School of Civil and Resource Engineering, The University of Western Australia, Crawley, WA, 6009, Australia (hao@civil.uwa.edu.au)

INTRODUCTION

Structural Health Monitoring (SHM) aims to assess, and where possible extend, the safe service life of bridges, buildings and other civil infrastructure. However, while SHM can improve predictions of damage and structural performance for existing structures, an assessment still needs to be made about how safe the existing structure is. Design codes are not useful for this purpose, as they are overly conservative and intended for design of new structures not safety assessment of existing structures. Site-specific information is generally available for the assessment of a specific structure, and this new (updated) information can often reduce uncertainties about load and resistance, leading to more accurate estimation of limit state accidence. For instance, AS5100-7 (2004) provides specific requirements for safety assessment of bridges and AS ISO 13822 (2005) for structural assessment in general. These assessment standards still define safety in a deterministic manner as when a factored (conservative) resistance exceeds a factored (conservative) load, as is the case with design codes. Owing to their conservative nature, deterministic safety assessment methods are not ideally suited to predicting structural performance, particularly if it involves progressive damage. While conservative assumptions at the time of design generally have little influence on the cost of new structures, their application to existing structures is highly likely to condemn them unnecessarily, since most structures have unrealized reserve capacity or performance, not indicated by design methods (Melchers, 2001). There is thus a need to consider all of the variability and uncertainty that affect structural resistance and load estimation. Reliability-based assessment has the capability to deliver estimates of the risks involved in the assessment and prediction processes, as well as maintenance and service life predictions and is well-suited to modern cost-benefit-risk decision procedures.

Reliability-based safety assessment is used frequently in Europe, and the Danish Roads Directorate (DRD). DRD is one of the few authorities to provide very specific guidelines on the reliability-based assessment of existing bridges (DRD, 2004). While deterministic safety assessments are appropriate for most bridge assessments, if an assessment recommends bridge closure, load restriction or extensive and costly strengthening, it is often useful to undertake a more detailed reliability-based assessment. For example, the DRD now pursues reliability-based assessment as a matter of course for all structures that have failed a deterministic assessment, and probability-based assessments on 11 bridges have saved the DRD over \$35 million (O'Connor and Enevoldsen, 2007). Note that Australian Standards now provide guidance on reliability-based assessment of existing structures (AS5104-2005, AS ISO 13822-2005), and work is progressing on applying structural reliability assessment to SHM applications (e.g., Wong and Yao, 2001; Necati et al., 2008; Dissanayake and Karunananda, 2008). The present paper will describe reliability-based assessment procedures and then present a spatial time-dependent structural reliability analysis to update the deterioration process and evaluate the probability of structural failure and service life prediction based on vibration results for a reinforced concrete (RC) corroding beam.

Periodic on-site inspections are normally conducted to assess the current condition and structural reliability of the structure, as well as reliability into the future, which will aid reliability-based service life predictions (e.g., Suo and Stewart, 2009). Corrosion damage can lead to a change in both static and dynamic behaviour (Cawley and Adams, 1979). For this reason, a useful damage-detection method (based on dynamic testing) may be one that monitors changes in resonant frequencies because frequency measurements can be quickly conducted, are non-destructive and are often given reliable results (Salawu, 1997). However, vibration test data can be contaminated by measurement errors (Capozucca, 2008), which is why Xia and Hao (2003) and others have developed statistical approaches to help deal with test data uncertainty to allow model updating and estimation of the probability of damage detection.

As the reinforcement corrosion process is often accompanied with cracking, spalling and de-bonding, their influence on dynamic response and structural stiffness is a complex phenomenon to model. While models exist to relate damage and changes in natural frequency (Unger et al., 2005; Zhu and Law, 2007), the ability to relate the spatial and time-dependent aspects of reinforcement corrosion, such as localised pitting corrosion, appears to be lacking. For example, an accurate assessment of structural capacity requires that the location and extent of corrosion damage is known. For this reason, several laboratory experiments have investigated the influence of reinforcement corrosion on structural dynamic characteristics (e.g., Capozucca and Cerri, 2000; Razak and Choi, 2001). Both of these studies involved only two small-scale RC beams, where it was observed that a decrease in natural frequency occurred as corrosion increased. However, Capozucca and Cerri (2000) did not estimate the corrosion loss, and Razak and Choi (2001) recorded corrosion losses of 7.7% for both beams. Hence, there is a need for the influence of corrosion loss on RC beam natural frequencies to be estimated for a range of corrosion losses.

It is of interest to update the safety and reliability assessment of RC structures in corrosive environments using natural frequency vibration (damage detection) results. This paper analyzes the experimental results of natural frequencies obtained from six RC beams subjected to accelerated chloride-induced corrosion where corrosion loss varies from 1 to 20% (Wang et al., 2008). These experimental results are used to establish an empirical model, which describes the relationship between natural frequency and corrosion loss. A statistical analysis of model error (accuracy) is used to characterize the variability of corrosion loss predictions. The reliability assessment of RC structures in corrosive environments can be updated if the natural frequency is known from a condition assessment. As only the first order frequency, which reflects the global response of the structures, was recorded and highly localized pitting corrosion is inevitable during the corrosion process, a spatial time-dependent reliability model that considers the pitting damage of reinforcing bars is used for the spatial time-dependent structural reliability analysis. The performance of RC beams is used to illustrate the reliability analysis developed in this paper. The inspection finding considered herein is the timing of test and the vibration results, which is used to provide an updated estimate of structural reliability. This reliability approach can be used for service life prediction and is a very powerful tool for SHM.

RELIABILITY-BASED ASSESSMENT OF EXISTING STRUCTURES

Structural Reliability

If the limit state of interest is related to structural capacity, then failure is deemed to occur when a load effect (S) exceeds structural resistance (R). The probability of failure (p_f) is

$$p_f = \Phi(-\beta) = \Pr(R \le S) = \int_0^\infty F_R(r)f_s(r)dr \tag{1}$$

where R and S are statistically independent random variables, $f_S(r)$ is the probability density function of the load, $F_R(r)$ is the cumulative probability density function of the resistance, $\Phi()$ is the standard Normal distribution function (zero mean, unit variance) extensively tabulated in statistics texts, and β is the "reliability index" or "safety index." In structural reliability literature, particularly reliability-based calibration of codes, the reliability index is often used as the measure of safety. The relationship between probability of failure p_f and reliability index β is shown in Table 1.

A limit state is a boundary between desired and undesired performance and is referred to as G(X), where the vector X represents the basic variables involved in the problem, which in the present case is equal to R-S. For many realistic problems, the simplified formulation given by Eqn. 1 is not sufficient, as usually several random variables will influence structural capacity, such as material properties, dimensions, model error, etc. Moreover, there are likely to be several load processes acting on the system at the same time. Equation 1 can be generalized to

$$p_f = \Pr[G(\mathbf{X}) \le 0] = \int \ldots \int_{G(\mathbf{X}) \le 0} f_\mathbf{X}(\mathbf{x})d\mathbf{x} \tag{2}$$

where $f_x(x)$ is the joint probability density function for the n-dimensional vector $X=\{X_1,....,X_n\}$ of random variables, each representing a resistance random variable or a loading random variable acting on the system. Note that failure probability may be calculated per annum, per lifetime or for any other time period. It follows that the probabilistic models selected for loads and resistances must relate to this time period. Figure 1 shows a representation of a joint probability density function for two variables R and S and the probability of limit state exceedance.

Table 1. Relationshiop Between Probability Of Failure (P_f) And Reliability Index (β)

p_f	β
10^{-1}	1.28
10^{-2}	2.33
10^{-3}	3.09
10^{-4}	3.71
10^{-5}	4.26
10^{-6}	4.75
10^{-7}	5.19
10^{-8}	5.62
10^{-9}	5.99
10^{-10}	6.36

The solution to Eqn. 2 is complex, particularly since resistance and loading variables often vary in time and space (such as corrosion, fatigue, or other sources of deterioration), the variables may be correlated, and most infrastructure comprises many elements or components requiring a systems approach to infrastructure performance. For example, resistance and load processes are spatial and time-dependent, and Figure 2 shows that resistance may decrease due to deterioration and that loading is a highly stochastic process.

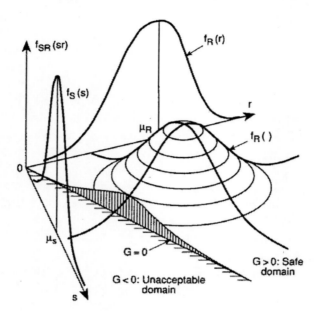

Figure 1. Region of Integration for Failure Probability (Melchers, 1999).

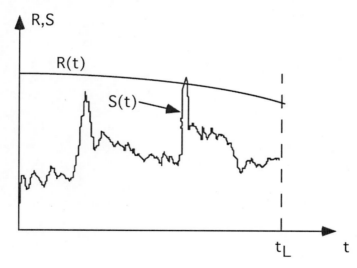

Figure 2. Resistance and Load Stochastic Processes

Two main approaches can be taken to solve Eqn. 2:

1. analytical methods by transforming $f_x(\mathbf{x})$ to a multi-normal probability density function to approximate the failure region of the limit state function—Second

Moment and transformation methods such as First Order Second Moment (FOSM) and First Order Reliability Methods (FORM).

2. numerical approximations to perform the multidimensional integration required in Equation 2—Monte-Carlo simulation techniques that involve random sampling each random variable, and the limit state G(X) is then checked, enabling p_f to be inferred from a large number of simulation runs.

These approaches are all not of equal accuracy or applicable to all problems. Second moment and transformation methods are computationally very efficient and often very useful for most problems. They do have disadvantages, however, that non-linear limit state functions are not easily handled and may give rise to inaccuracies and difficulties can arise in using non-normal random variables and dependencies. On the other hand, simulation methods are, in principle, very accurate and can handle any form of limit state function and are not restricted to normal random variables. However, the computational times can be significant, due to the high number of simulation runs often needed to produce convergent results. Nonetheless, with the availability of computers with ever-increasing speed, and the use of importance sampling, response surface methods and other variance reduction techniques, the computational efficiency of simulation methods can be greatly improved.

The probabilistic description of resistance and load variables can only be assumed in the design stage; however, parameter variability is usually obtained from the collection and statistical analysis of data from existing structures. This is the basis for reliability-based design code calibration (e.g., Ellingwood et al., 1980; Pham 1985; Pham and Bridge, 1985; Nowak et al., 2005). This approach is suitable for a whole class of structures, but not for assessing the safety of individual structures. In this case, the probabilistic models for resistance and load variables can be updated using site-specific data, such as testing of concrete cores or steel coupons, cover meter measurements, weigh in motion (WIM) traffic data, service proven performance and other condition assessment data. For example, as yield strength is a highly variable parameter in structural steel bridges, it is often recommended that tests or samples be taken from the assessed structure to better characterize the yield strengths for the structure under consideration (e.g., Diamantidis 2001). The incorporation of test data into a Bayesian statistical framework often results in the selection of parameters with lower variability and so leads to higher estimates of structural reliability. SHM data is additional condition assessment data that will further improve predictions of structural damage and capacity, which will lead to improved estimates of structural reliability. The impact of Bayesian updating on structural reliability predictions can be considerable, often leading to increases in structural reliability (Stewart and Val, 1999; Val et al., 2000).

Target Reliability and Service Life Prediction

For structural safety assessment, a safe service life is defined to be exceeded when the probability of failure exceeds a target reliability. The Australian and ISO Standards "General Principles on Reliability of Structures" (AS5104-2005, ISO 2394-1998) suggest that the lifetime target reliability index (β_T) is 3.1 to 4.3 for ultimate (strength) limit states design (see Table 2). In structures where there is little redundancy, a higher target reliability index may be selected. Such target values are "informative" only, as the selection of the target reliability level depends on the different parameters such as type and importance of the structure, possible failure consequences, socio-economic criteria, etc. (e.g., Diamantidis, 2001; Nowak and Collins, 2000).

Table 2. Suggested Lifetime Target Reliabilities (β_T) (As5104-2005)

Relative Costs of Safety Measures	Consequences of Failure			
	Small	Some	Moderate	Great
High	0	1.5	2.3	3.1
Moderate	1.3	2.3	3.1	3.8
Low	2.3	3.1	3.8	4.3

The Danish Road Directorate provides guidelines for reliability assessment of existing bridges for an ultimate limit state with a high safety class (see Table 3). Note that target reliabilities are for annual probabilities of failure and assume that probabilistic models similar to those described in the guide will be used. The target failure probability for t years $P_{ft}(t)$ is

$$P_{ft}(t) = 1 - (1 - P_{fA})^t \qquad (3)$$

where P_{fA} is the annual probability of failure and assuming statistically independent events. This means that a longer planned service life will result in a larger target reliability.

Table 3. Annual Target Reliabilities For Ultimate Limit States (Drd, 2004)

Failure type	Failure with warning and bearing capacity reserve	Failure with warning but without reserve capacity	Failure without warning
β_T	4.26	4.75	5.20
p_{fA}	10^{-5}	10^{-6}	10^{-7}

Although the emphasis of the chapter is on structural engineering systems, the computational and probabilistic methods described herein are also appropriate for other load-resistance or demand-capacity systems, such as geotechnical, mechanical, hydraulic, electrical and electronic systems where performance failure is defined as when a predicted load/demand exceeds a resistance/capacity. For more details on structural reliability theory, see Ditlevsen and Madsen (1996), Melchers (1999), and Nowak and Collins (2000).

Finally, structural reliability methods can provide valuable predictive capability for optimizing life-cycle costs. In this case, probability of failure estimates are used to calculate expected damage/failure costs for various design and maintenance interventions, and so their influence on life-cycle costs can be assessed (e.g., Val and Stewart, 2003). This approach has significant utility to SHM where although initial costs may be higher, their effect on increasing structural reliability may well result in reduced life-cycle costs when they are assessed over the 50- to 100-year service life of a structure.

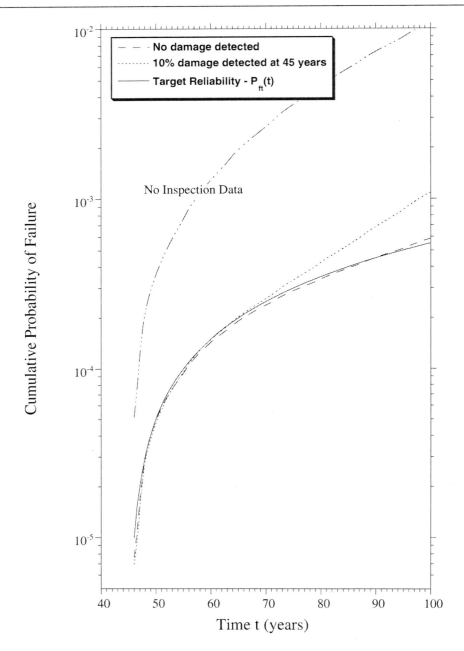

Figure 3. Reliability Updating Based on Visual Inspection Data (Stewart, 2009a)

Example - Rc Corroding Beam

To illustrate the utility of reliability-based safety assessment, assume that a RC beam exposed to an aggressive chloride environment has survived for 45 years. Visual inspections of corrosion damage (chloride-induced cover cracking) revealed no corrosion damage at 42 years, but at 45 years, another inspection reveals 10% of the beam's surface suffers corrosion damage. A spatial time-dependent reliability analysis is used to calculate updated probabilities of failure (structural collapse due to pitting corrosion) and compare this with the target reliability—for full details, see Stewart (2009a). The probabilities of failure p_f (t) are conditional on this inspection scenario, and another assuming that no damage is observed at either inspection, are shown in Figure 3. By means of comparison, the probability of failure considering no inspection data is also shown in Figure 3. The benefits of reliability updating, in terms of increasing structural reliability, are immediately apparent from Figure 3, where inspection data results in updated probabilities of failure that are at least an order of magnitude lower than if updating is not considered. This is to be expected, as there is high variability associated with deterioration and prediction models used in the time-dependent and spatial models, thus leading to high variability of the occurrence and extent of corrosion damage. The variability of predictions can be minimized, and confidence of results would be improved if results are updated with visual inspection or other condition assessment data.

Figure 3 also shows the target failure probability (P_{ft}), which as observed from Eqn. 3 will vary with time. The target reliability is based on the Danish Road Directorate guidelines, where for ductile mechanical behaviour of a RC beam, the annual failure probability $P_{fA}=10^{-5}$ for collapse with warning is appropriate (DRD, 2004). Clearly, if no inspection data are used, then the target failure probability is exceeded at the time of the second inspection (45 years). However, if no corrosion damage is observed, and this new information is used to update estimates of structural reliability, then the target failure probability is not exceeded until 89 years—thus extending the predicted safe service life by 44 years. This is a significant extension of service life prediction. Even if some corrosion damage is detected during an inspection, then the predicted safe service life is still relatively high at 64 years. These examples illustrate the benefits of inspection and condition assessment updating on service-life prediction.

EXPERIMENTAL PROGRAMME FOR EFFECT OF CORROSION-INDUCED DAMAGE ON VIBRATION TEST RESULTS

Experimental Set-Up

Six RC rectangular beams of 250×300 mm with a length of 1.5 m were fabricated at The University of Western Australia. In each beam, two 10 mm diameter plain round reinforcing bars are fabricated in the concrete matrix (see Figure 4). The concrete cover was 50 mm, the cement used for all specimens was ordinary Portland cement, and the water-cement ratio was 0.55 for all specimens. The coarse and fine aggregates (20 and 10 mm, respectively) were kept the same for all mixtures. Three percent of calcium chloride ($CaCl_2$) by weight of cement was added to the concrete mixture to induce corrosion along the reinforcing bars. This admixture had a negligible effect of concrete strength properties. All the specimens were moist-cured for 28 days before testing. Standard cylinders were tested at 28 days to determine the mean concrete compressive strength of 39.70 MPa, Young's modulus of 29.61GPa, and concrete mass density of 2426.75 kg/m^3.

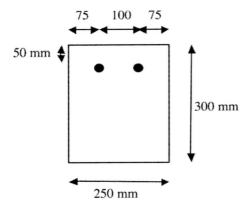

Figure 4. Cross-section of RC Specimens.

In order to generate corrosion damage within a reasonable time period, an accelerated nominal corrosion rate 100 μA/cm^2 was used in this study. A schematic of the experimental set-up for the accelerated corrosion testing is shown in Figure 5. The soffit of the specimen was immersed in a 5% Sodium Chloride (NaCl) solution. The accelerated corrosion process was achieved by applying a constant electrical current to the bars via a current regulator. The steel reinforcing bar acts as the anode, the stainless steel plate submerged in the NaCl solution acts as the cathode, and the pore fluid in the concrete acts as the electrolyte.

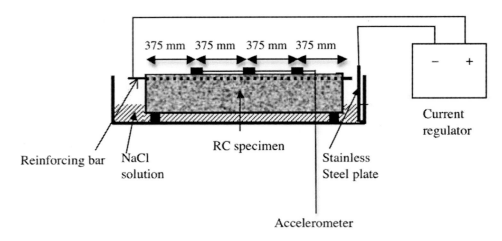

Figure 5. Experimental Setup of Accelerated Corrosion Test.

The tests lasted for up to 24 weeks. The time to crack initiation and propagation on the concrete surface were verified by visual observations using a crack detector microscope with an accuracy of 0.05 mm twice a week. The number, length and maximum width of cracking corresponding to each bar were recorded.

The first modal test was conducted when the beams were first placed in the NaCl solution and before any external current was applied. After that, dynamic tests were taken to obtain modal parameters of each beam every two weeks. An impact hammer was used to give excitation to the beams. Because only the first mode was used in this study, three measurement points were taken as shown in Figure 5. The first natural frequency and corresponding damping ratio were then obtained. After every four weeks, one specimen was removed from testing in order to accurately determine the amount of corrosion at the end of the testing period. The gravimetric weight loss method as specified in ASTM G1-90 (1990) was used. At the completion of the accelerated testing, the reinforcing bars were removed, cleaned, and the weight loss of the bars was measured. Assuming general corrosion, the measured corrosion rate was calculated for each specimen.

Experimental Results

During the accelerated corrosion tests, crack widths of up to 7 mm were observed to occur above and parallel to the reinforcing bars. Even after four weeks, the crack width was near 1 mm, and increased in a linear function of approximately 1 mm per month. For more details of crack width measurements, see Wang et al. (2008).

The gravimetric weight loss method was used to calculate corrosion loss (Q_{corr}), defined herein as the percentage loss of reinforcement cross-section. The rates of corrosion loss (Q_{corr}/day) for the six RC specimens (average of two rebars) are calculated by dividing the final corrosion loss by the time period, and this is shown in Table 4. It is clear to see that there is high variability in the corrosion rate even under the same experimental environment. The coefficient of variation of corrosion loss is calculated as COV = 0.126. The corrosion loss at the time of dynamic testing can be calculated by multiplying the rate of corrosion loss by the time of dynamic testing.

Table 4. Experimental Values of Corrosion Loss

Specimen	1	2	3	4	5	6
Corrosion Rate (Q_{corr}/day)	0.120%	0.099%	0.137%	0.108%	0.135%	0.139%

Figure 6 shows the experimental results for natural frequency (f) and corrosion loss (Q_{corr}). There is a clear trend that natural frequency reduces as corrosion loss increases. It is not unexpected that when a RC beam is subjected to corrosion damage, its stiffness will change due to the reduction of rebar cross-sectional area, de-bonding, cover cracking and delamination, which will consequently influence natural frequency. The trends shown in Figure 6 are consistent with the experimental observations by Capozucca and Cerri (2000) and Razak and Choi (2001).

An empirical regression equation that describes the relationship between corrosion loss and measured frequencies (see Figure 6) is

$$Q_{corr} = 0.3 + 3.13 \times 10^{12} e^{-0.0591f} \tag{4}$$

where f is the tested natural frequency of the beam (Hz). The magnitude of the frequency changes is an indicator of the severity of corrosion damage. The safety and reliability assessment of RC structures in a deteriorating environment can then be updated by using such natural frequency results. The probabilistic characterisation of model error (ME) is needed for any reliability analysis where model error is the ratio of actual (experimental) value to predicted value. In this case, a statistical analysis of the comparison of experimental data with Eqn. 4 yields the following statistical parameters: mean (ME)=1.0 and σ(ME)=1.67%.

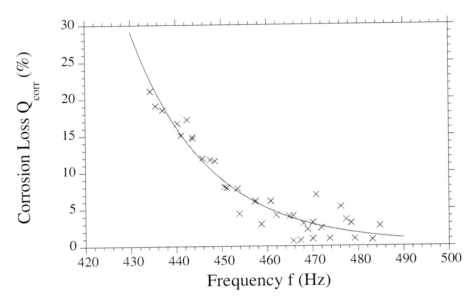

Figure 6. Empirical Model for Prediction of Corrosion Loss Based on Frequency Measurements.

An assessment of strength and reliability requires that corrosion loss can be converted to corrosion rate, pit depth and associated time-dependent loss of reinforcement cross-sectional area. For a constant corrosion rate with time, the corrosion current density i_{corr} can be inferred from corrosion loss Q_{corr} at time of test t_{test} (years) as

$$i_{corr} = \frac{D_o\left(1 - \sqrt{1 - \dfrac{Q_{corr}(t_{test})}{100}}\right)}{0.0232 t_{test}} \qquad (5)$$

where D_o is initial diameter of reinforcing bar (mm) and corrosion rate i_{corr} is $\mu A/cm^2$. Faraday's Law indicates that a corrosion current density of $i_{corr}=1\ \mu A/cm^2$ corresponds to a uniform steel section loss of 0.0116 mm/yr, hence the maximum pit depth along a reinforcing bar can be estimated as

$$p(t) = 0.0116 \times R \times i_{corr} \times t \qquad (6)$$

where t is the time since corrosion initiation, and R is the pitting factor, which is equal to the ratio of maximum pitting depth p(t) to corrosion penetration due to general corrosion (p_{av}). Experimental results show that R is statistically independent for 100 mm lengths and the Gumbel distribution is the best fit for modelling maximum pit depth for reinforcing bars (Stewart and Al-Harthy, 2008). The statistical parameters for 100 mm reinforcing steel with

10 mm diameter are mean(R)= 5.65 and COV(R)=0.22 (Stewart and Al-Harthy, 2008). To predict the distribution of maximum pitting factor of any length, the Gumbel statistical parameters can be modified as described by Stewart (2004). The pitting depth p(t) can be used to calculate the loss area $A_{pit}(t)$, and the corresponding remaining area of cross-section of a reinforcing bar due to pitting corrosion by the geometric model given by Val and Melchers (1997).

Du et al. (2005) have observed that, due to local attack penetration, the residual yield and ultimate strength of reinforcement decreases significantly, and the yield strength f_y reduces linearly with corrosion loss such that

$$f_y(t) = f_{yo}\left[1.0 - \alpha_y Q_{corr}(t)\right] \tag{7}$$

where f_{yo} is the yield stress of an uncorroded reinforcing bar and α_y is an empirical coefficient (=0.005).

SPATIAL TIME-DEPENDENT RELIABILITY ANALYSIS

It is well accepted that many variables that influence the corrosion process and structural capacity, such as the concrete strength, cover, environment, etc., are not homogeneous across the whole structure. In other words, there is temporal and spatial variability of the corrosion process and resulting structural capacity. For example, Stewart and Mullard (2007), Stewart (2009b) and others have shown that pitting corrosion is a spatial and temporal variable, and random field theory and spatial time-dependent reliability analyses can be used to predict corrosion damage and loss of structural safety and reliability for deteriorating structures. Stochastic models that do not include random spatial variability cannot adequately describe the damage and performance characteristics of many RC structures.

A spatial time-dependent reliability analysis based on random field theory is a necessary and useful tool to predict the likelihood and extent of cracking for RC structure surfaces (Vu and Stewart, 2005; Stewart and Mullard, 2007; Mullard and Stewart, 2009), as well as pitting corrosion for RC and prestressed concrete structures (Darmawan and Stewart, 2007; Stewart and Suo, 2009). According to random field theory, the structure is divided into m elements of equal size, and the random variables within the random field are statistically correlated. For the ease of computation, the middle point method is adopted. The value of each element, such as concrete compressive strength, will be represented by the values of the centroid of each element, and this value is assumed to be constant within the element. Elements in a random field are usually spatially correlated. The Gaussian correlation function is used in this paper to calculate the correlation coefficient between any two elements for concrete compressive strength. The scale of fluctuation is taken as 2.0 m (Stewart and Mullard, 2007). The pitting factor (R) is statistically independent between elements; this is equivalent to a random field with a scale of fluctuation of zero. For more details about random field theory, see Vanmarcke (1983).

For the ultimate strength limit state, a RC beam composing of m elements can be treated as a series system. As failure of the beam corresponds to failure of the weakest element in the beam, the limit state occurs when load effects $S_j(t)$ exceed resistance $R_j(t)$ at any element at time t (Stewart, 2004)

$$G(t) = \min_{j=1,m}\left[R_j(t) - S_j(t)\right] \tag{8}$$

where $R_j(t)$ is the resistance for element j, and $S_j(t)$ is the load effects at the mid-point of element j. For example, the flexural loading capacity of element j of a singly reinforced RC beam is based on layout of n main reinforcing bars, which comprise a parallel system. If only the ductile mechanical behaviour of each reinforcing bar is considered, the time-varying flexural resistance of the jth element of the beam is (Stewart, 2009b)

$$R_j(t) = ME\sum_{i=1}^{n}A_i(t)f_{yi}(t)(d - \frac{\sum_{i=1}^{n}A_i(t)f_{yi}(t)}{1.7f_cb}) \tag{9}$$

where ME is model error for flexure, f_c is the concrete compressive strength, d is the effective depth of cross-section, b is the beam width, $A_i(t)$ is the cross-sectional area of the ith reinforcing bar and $f_{yi}(t)$ is the yield strength for the ith reinforcing bar. Clearly, corrosion rate and pitting factor will affect $A_i(t)$ and $f_{yi}(t)$.

Assume that a condition assessment (in this case, a dynamic test) is conducted at time t_{test}, and that K load events occur within the time interval (0, t_{test}) at times t_i (i = 1, 2,. . .,K) and that $t_K \leq t_{test}$. After the condition assessment, N load events occur within the time interval (t_{test},t). The result from the dynamic test is expressed as H. Service-proven performance may be viewed as the structure surviving a series of service loads which act as "proof" loads (Stewart, 1997). If the structure being inspected at time t_{test} has survived t_{test} years of loading up to this time, then the updated probability that a structure will fail in t years considering t_{test} years of service proven performance can be expressed as:

$$p_f(t) = \Pr\left[\{G(t_{n1}) < 0 \cup G(t_{n2}) < 0 \cup ... \cup G(t_N) < 0\}|\{G(t_1) > 0 \cap G(t_2) > 0 \cap ... \cap G(t_K) > 0\}\right]$$

$$\text{where } t_K \leq t_{test}, \ t_1 < t_2 < ... < t_K, \ t \geq t_{test}, \ t > t_{n1} > t_{n2} > ... > t_{test} \tag{10}$$

Equation (10) may be referred to as a hazard function or hazard rate. As resistance degradation includes a large number of random variables and spatial random fields, then Monte Carlo simulation is a powerful tool to solve Eqn. 10.

RESULTS

Structural Configuration

Since the empirical relationship between corrosion loss and frequency is appropriate only for the test specimen, the relevant analysis will be conducted for the test beam, assuming that it is supporting stochastic floor loading in proportion to its structural capacity. The design condition is $\phi R_{nom}=1.2G_n+1.6Q_n$ (ACI318-2005) where G_n and Q_n are design dead and live loads. In the present case, $\phi=0.9$, $Q_n/G_n=0.8$, and the load is uniformly distributed. The structural configuration of the test beam is b=250 mm, d=250 mm, concrete cover = 50 mm, two 10 mm diameter reinforcing bars each with nominal yield strength $f_{ynom}=400$ MPa, and mean concrete compressive strength (f_c) is 39.7 MPa. All variables are assumed as deterministic (as was the case for constructed test specimens), with the exception of yield strength, concrete strength and pitting factor, which for the test specimens are variable (in time or space). Statistical parameters for the stochastic loads are summarized in Table 5, and statistical parameters for material properties and corrosion parameters are shown in Table 6. As we are not modelling time to corrosion initiation, then corrosion starts at t=0. The 1.5 m RC beam is divided into M=6 elements with equal length of Δ=250 mm, see Figure 7. A 1D random field is used to model the spatial variability of concrete compressive strength and pitting factor of the six elements. For example, Figure 8 shows a typical simulation realization of pitting factors (R) for a 10 m RC beam with six 27 mm diameter bars, each with an element length of Δ=250 mm. The service life considered herein is 100 years.

Table 5. Statistical Parameters For Stochastic Load Model

Load	Duration	Mean	COV	Distribution	Reference
Dead Load	Permanent	$1.05G_n$	0.10	Normal	Ellingwood et al. (1980)
Live Load:					
Sustained	8 years	$0.30Q_n$	0.60	Gamma	Ellingwood and Culver (1977)
					Chalk and Corotis (1980)
Extraordinary	1 year	$0.19Q_n$	0.66	Gamma	Philpot et al. (1993)

Note: G_n, Q_n design loads specified from ANSI/ASCE 7-93.

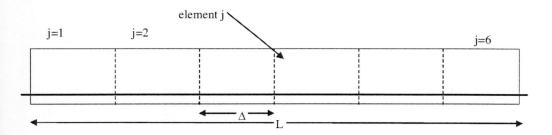

Figure 7. Discretisation of RC Beam.

Table 6. Statistical Parameters for Material and Deterioration Variables

Parameter	Mean	COV	Distribution
f_{yo}	$1.18 f_{ynom}$	0.04	Normal
Q_{corr}	1.0	0.126	Normal
$ME(Q_{corr})$	1.0	$\sigma = 1.67\%$	Normal
Compressive strength f_c	39.7 MPa	$\sigma = 6$ MPa	Normal
ME(flexure)	1.01	0.046	Normal
Pitting factor R (L= 250 mm)	6.55	0.19	Gumbel

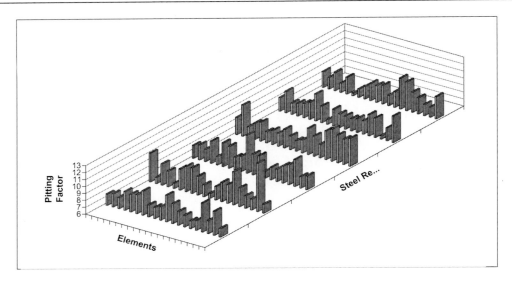

Figure 8. Monte-Carlo simulation realization of pitting factors (R) for six Y27 reinforcing bars for a 10 m RC beam discretised into 20 elements.

Dynamic Test Scenarios and Computational Method

Nine dynamic test scenarios are assumed for time of test (t_{test}) of 5, 10 or 35 years with test frequency results of 440, 460 and 480 Hz (see Table 7). The corrosion losses corresponding to the measured frequency are obtained from Eqn. 4, and the corrosion rate is then inferred from Eqn. 4. In this analysis, we assume that corrosion rate is constant with time and the same for all elements. The pitting factor and concrete compressive strengths are then randomly generated for each element. As the RC beam is discretised into m elements, then m values of resistance (R_j) and load effect (S_j) are obtained for each time step during each simulation run. In this case, corrosion loss matches the inspection data (H). As the time steps proceed past the time of inspection, then the simulated time-dependent damage progression is used to assess if the ultimate strength limit state is exceeded. After many simulation runs, it is then possible to infer the updated probability of failure given by Eqn. 10.

Table 7. Dynamic Test Scenarios And Corrosion Rates

Time of Test t_{test}	Test Scenario	Frequency (Hz)	Corrosion Loss Q_{corr}	Corrosion Rate i_{corr} ($\mu A/cm^2$)
5 years	1	440	16.22%	7.25
	2	460	5.18%	2.25
	3	480	1.80%	0.77
10 years	4	440	16.22%	3.62
	5	460	5.18%	1.12
	6	480	1.80%	0.39
35 years	7	440	16.22%	1.04
	8	460	5.18%	0.32
	9	480	1.80%	0.11

Results

The mean corrosion loss and mean corrosion rate for each dynamic test scenario are given in Table 7. While the predicted corrosion loss is the same irrespective of the time of test, the corrosion rate reduces as the time of test increases. This is expected, as the analysis assumes that corrosion starts immediately (t=0), so a larger time equates to a lower average corrosion rate. The "medium corrosion intensity" for a RC structure in an aggressive chloride environment is typically 1 $\mu A/cm^2$ (Dhir et al., 1994). Hence, test scenarios 3, 5 and 7 represent perhaps the most realistic test scenarios, whereas the other scenarios result in either very low or very high corrosion rates. The target reliability is based on the Danish Road Directorate guidelines for reliability assessment of existing bridges, where for ductile mechanical behaviour of a RC beam, the annual failure probability $P_{fA}=10^{-5}$ for collapse with warning is appropriate (DRD, 2004). The time-dependent change in target reliability is given by Eqn. 3.

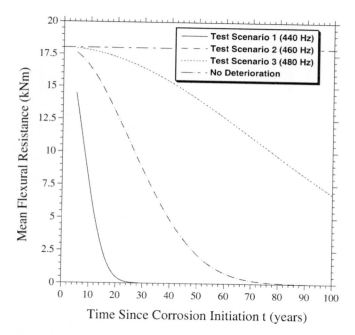

Figure 9a. Time-Dependent Resistance (t_{test}=5 years).

Figures 9a and 9b show the mean time-dependent structural resistance and probability of failure for test scenarios one to three. The high corrosion rates associated with test scenarios one and two result in very low structural resistance. Clearly, this will cause a high probability of failure, as shown in Figure 9b. Figure 9b shows that for test scenario two, the probability of failure exceeds the target reliability given by Eqn. 3 at t=13 years—this means that the service life prediction based on dynamic test results obtained from test scenario two (460 Hz) is 13 years. However, if 480 Hz is measured from the dynamic tests (test scenario three), then service life prediction is increased to 40 years. Figures 10 and 11 show similar trends for test scenarios four to nine, where tests are conducted at t_{test}=10 years and t_{test}=35 years, respectively. Note that because the corrosion loss is the same for the same measured vibration result, then structural resistance is identical and so probability of failure also very similar. The timing of the dynamic test results, however, affects the time-dependent nature of loss of structural resistance and reliability. For example, Figure 11b shows the service-life prediction based on dynamic test results obtained from test scenario 8 (460 Hz) is 96 years, but a same test result after only ten years (Figure 10b) produces a reduced service life prediction of only 27 years.

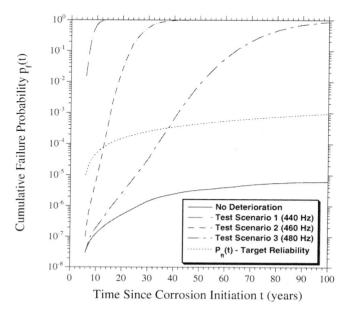

Figure 9b.Time-Dependent Probabilities of Failure (t_{test}=5 years).

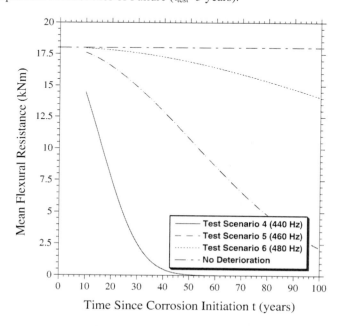

Figure 10a. Time-Dependent Resistance (t_{test}=10 years).

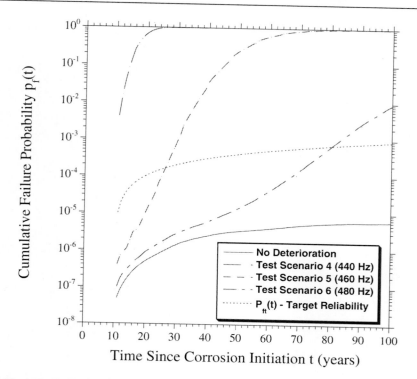

Figure 10b. Time-Dependent Probabilities of Failure (t_{test}=10 years).

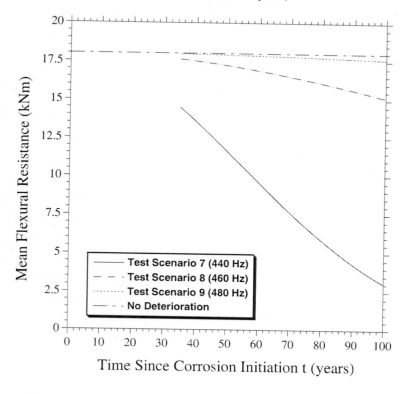

Figure 11a. Time-Dependent Resistance (t_{test}=35 years).

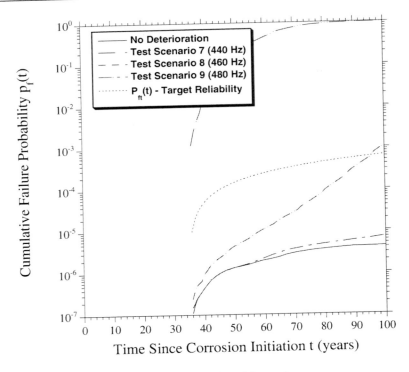

Figure 11b. Time-Dependent Probabilities of Failure (t_{test}=35 years).

More test scenarios can be considered, but those used herein simply help illustrate that the time-dependent structural performance and reliability can be updated with information after dynamic test results. The updated and more accurate structural reliability predictions containing latest condition assessment information can be used to improve engineering management strategies such as inspection and maintenance schedules and costs, as well as more accurate service life prediction.

CONCLUSION

The utility of reliability-based safety assessment for corroding structures has been illustrated where updated estimates of structural reliability based on condition assessment data can often lead to increases in safe service lives. Accelerated corrosion tests were conducted on six RC specimens to assess the changes of dynamic characteristics due to reinforcement corrosion. An empirical model is developed to describe the relationship between corrosion loss and frequency. A spatial time-dependent reliability model is developed to model corrosion rate, pitting, loss of structural capacity and probability of failure conditional on dynamic test results. The spatial time-dependent reliability model includes the temporal and spatial variability of concrete strength and pitting corrosion. Nine dynamic testing scenarios are assumed based on different vibration results at time of test from five to 35 years. It was found that dynamic test results change future reliability predictions of structural reliability significantly. The estimates of structural reliability can be compared with target reliabilities for improved service life prediction.

ACKNOWLEDGMENTS

The support of the Australian Research Council and the Cooperative Research Centre for Integrated Engineering Asset Management (ID207) are gratefully acknowledged. The experimental tests were conducted at The University of Western Australia, and the authors appreciate the assistance of Dr. Xinqun Zhu and Mr. Ying Wang. The assistance of Dr. Mukshed Ahammed with computer programming is also gratefully acknowledged.

REFERENCES

ACI 318, *Building Code Requirements for Structural Concrete*, ACI, Detroit, Michigan, 2005.

AS5104-2005, *General Principles on Reliability for Structures*, Standards Australia, Sydney.

AS5100-7 (2004*), Bridge Design - Rating of Existing Bridges*, Standards Australia, Sydney.

AS ISO 13822 (2005), *Basis for Design of Structures - Assessment of Existing Structures*, Standards Australia, Sydney.

ASTM G1-90 (1990). Standard Practise for Preparing, Cleaning and Evaluating Corrosion Test Specimens. *American Society for Testing and Materials*. Philadelphia, PA.

Cawley, P. and Adams, R.D. (1979), "The location of Defects in Structures from Measurements of Natural Frequencies." *Journal of Strain Analysis*, 14(2) pp. 49-57.

Capozucca R. (2008), "Detection of Damage Due to Corrosion in Prestressed RC Beams by Static and Dynamic Tests." *Construction and Building Materials*, 22(5) pp. 738-746.

Capozucca, R. and Cerri, M.N. (2000), "Identification of Damage in Reinforced Concrete Beams Subjected to Corrosion." *ACI Structural Journal*, 97(6) pp. 902-909.

Chalk, P.L. and Corotis, R.B. (1980), "Probability Models for Design Live Loads." ASCE *Journal of the Structural Division*, 106(ST10), pp. 2017-2033.

Darmawan, M.S., and Stewart, M.G. (2007), "Spatial time-dependent reliability analysis of corroding pretensioned prestressed concrete bridge girders." *Structural Safety*, 29(1) pp. 16-31.

Dhir, R.K., Jones, M.R., and McCarthy, M.J. (1994), "PFA concrete: chloride-induced reinforcement corrosion." *Magazine of Concrete Research*, 46(169) pp. 269–77.

Dissanayake, P.B.R. and Karunananda, P.A.K. (2008), "Reliability Index for Structural Health Monitoring of Aging Bridges." *Structural Health Monitoring*, 7(2) pp. 175-183.

Diamantidis, A. (2001), Probabilistic Assessment of Existing Structures, *A Publication of the Joint Committee on Structural Safety* (JCSS), RILEM Publications S.A.R.L.

Ditlevsen, O. and Madsen, H.O. (1996), *Structural Reliability Methods*, Wiley, Chichester, UK.

DRD (2004), Reliability-based classification of the load carrying capacity of existing bridges. *Denmark Guideline Document*, Report 291, 2004

Du, Y.G., Clark, L.A., and Chan, A.H.C. (2005), "Residual Capacity of Corroded Reinforcing Bars." *Magazine of Concrete Research*, 57(3) pp. 135-147.

Ellingwood, B., Galambos, T.V., MacGregor, J.G. and Cornell, C.A. (1980*), Development of a Probability Based Load Criterion for American National Standard A58, National*

Bureau of Standards Special Publication 577, U.S. Government Printing Office, Washington D.C.

Ellingwood, B. R. and Culver, C. G. (1977), "Analysis of Live loads in Office Buildings." *Journal of the Structural Division*, ASCE, 103(ST8) pp. 1551-1560.

ISO 2394 (1998*), General Principles on Reliability for Structures, International Organization for Standardization*, Geneva.

Melchers, R.E. (1999), Structural Reliability Analysis and Prediction, Wiley, Chichester, UK.

Melchers, RE (2001) "Assessment of Existing Structures—Some Approaches and Research Needs." *Journal of Structural Engineering*, ASCE, 127(4) pp. 406-411.

Mullard, J.A. and Stewart, M.G. (2009), "Stochastic Assessment of the Timing and Efficiency of Maintenance for Corroding RC Structures." *Journal of Structural Engineering*, ASCE, 135(8) pp. 887-895.

Necati, C.F., Melih, S. and Frangopol, D.M. (2008), "Structural Health Monitoring and Reliability Estimation: Long-span Truss Bridge Application with Environmental Monitoring Data." *Engineering Structures*, 30(9) pp. 2347-2359.

Nowak, A.S. and Collins, K.R. (2000), Reliability of Structures, McGraw-Hill, Boston.

Nowak, A.S., Szerszen, M.M., Szeliga, E.K., Szwed, A. and Podhorecki, P.J. (2005), *Reliability-Based Calibration for Structural Concrete*, Report No. UNLCE 05-03, Department of Civil Engineering, University of Nebraska.

O'Connor, A. and Enevoldsen, I. (2007), "Probability-based Bridge Assessment." *Proceedings of the Institution of Civil Engineers*, 160(3) pp. 129-137.

Pham, L. and Bridge, R.Q. (1985*), "Safety Indices for Steel Beams and Columns Designed to AS1250-1981."* Civil Engineering Transactions, IEAust, CE27(1) pp. 105-110.

Pham, L. (1985), *"Load Combinations and Probabilistic Load Models in Limit State Codes."* Civil Engineering Transactions, IEAust, CE27(1) pp. 62-67.

Philpot, T.A., Rosowsky, D.V., and Fridley, K.J. (1993), "Serviceability Design in LRFD for Wood." *Journal of Structural Engineering*, 119(12) pp. 3649-3667.

Razak, H.A. and Choi, F.C. (2001), "The Effect of Corrosion on the Natural Frequency and Model Damping of Reinforced Concrete Beams." *Engineering Structures*, 23(9) pp. 1126-1133.

Salawu, O.S. (1997), "Detection of Structural Damage through Change in Frequency: A Review." *Engineering Structures*, 19(9) pp. 718-723.

Stewart, M.G. (1997), "Time-Dependent Reliability of Existing RC Structures." *Journal of Structural Engineering*, ASCE, 123(7) pp. 896-903.

Stewart, M.G. and Val, D. (1999), "Role of Load History in Reliability-Based Decision Analysis of Ageing Bridges." *Journal of Structural Engineering*, ASCE, 125(7) pp. 776-783.

Stewart, M.G. (2004), "Spatial Variability of Pitting Corrosion and its Influence on Structural Fragility and Reliability of RC Beams in Flexure." *Structural Safety*, 26(4) pp. 453-470.

Stewart, M.G. and Mullard, J.A. (2007), "Spatial Time-dependent Reliability Analysis of Corrosion Damage and the Timing of First Repair for RC Structures." *Engineering Structures*, 29(7) pp. 1457-1464.

Stewart MG, Al-Harthy A. (2008), "Pitting Corrosion and Structural Reliability of Corroding RC Structures: Experimental Data and Probabilistic Analysis." *Reliability Engineering and System Safety*, 93(3) pp. 273-382.

Stewart, M.G. (2009a), "Strength, Reliability and Asset Management of Corroding RC Structures." *Corrosion & Prevention* 2009, Coffs Harbour, 15-18 Nov. 2009.

Stewart, M.G. (2009b), "Mechanical Behaviour of Pitting Corrosion of Flexural and Shear Reinforcement and its Effect on Structural Reliability of Corroding RC Beams." *Structural Safety*, 31(1) pp. 19-30.

Stewart, M.G. and Suo, Q. (2009), "Extent of Spatially Variable Corrosion Damage as an Indicator of Strength and Time-dependent Reliability of RC beams." *Engineering Structures*, 31(1) pp. 198-207.

Suo, Q. and Stewart, M.G. (2009), "Corrosion Cracking Prediction Updating of Deteriorating RC Structures Using Inspection Information." *Reliability Engineering and System Safety*, 94(8) pp. 1340-1348.

Unger, J.F., Teughels, A. and De Roeck, G. (2005), "Damage Detection of a Prestressed Concrete Beam Using Modal Strains." *Journal of Structural Engineering*, ASCE, 131(9) pp. 1456-1463.

Val, D.V. and Melchers, R.E. (1997), "Reliability of Deteriorating RC Slab Bridge/" *Journal of Structural Engineering*, 123(12) pp. 1638-1644.

Val, D., Stewart, M.G. and Melchers, R.E. (2000), "Life-Cycle Performance of Reinforced Concrete Bridges: Probabilistic Approach." *Journal of Computer Aided Civil and Infrastructure Engineering*, 15(1) pp. 14-25.

Val, D.V. and Stewart, M.G. (2003), "Life Cycle Cost Analysis of Reinforced Concrete Structures in Marine Environments." *Structural Safety*, 25(4) pp. 343-362.

Vanmarcke, E.H. (1983), *Random field: analysis and synthesis*. Cambridge: The MIT Press.

Vu, K.A.T. and Stewart, M.G. (2005), "Predicting the Likelihood and Extent of Reinforcement Concrete Corrosion-induced Cracking." *Journal of Structural Engineering*, 131(11) pp. 1681-1689.

Wang, Y., Zhu, X., Hao, H. and Stewart, M.G. (2008), *"Corrosion-induced Cracking of Reinforced Concrete Beam: Experimental Study"* Proceeding of the 3rd Word Congress on Engineering Asset Management and Intelligent Maintenance System (WCEA-IMS 2008), Beijing China, pp. 1741-1746.

Wong, F.S. and Yao, J.T.P. (2001), "Health Monitoring and Structural Reliability as a Value Chain." *Computer-Aided Civil and Infrastructure Engineering*, 16(1) pp. 71-78.

Xia, Y. and Hao, H. (2003), "Statistical Damage Identification of Structures with Frequency Changes." *Journal of Wind and Vibration*, 263 pp. 853-870.

Zhu, X.Q. and Law, S.S. (2007), *"Damage Detection in Simply Supported Concrete Bridge Structure Under Moving Vehicular Loads."* Transactions of the ASME, 129(Feb) pp. 58-65.

In: Structural Health Monitoring in Australia
Editors: Tommy H.T. Chan and D. P. Thambiratnam

ISBN: 978-1-61728-860-9
©2011 Nova Science Publishers, Inc.

Chapter 8

Improving the Reliability of a Bridge Deterioration Model for a Decision Support System Using Artificial Intelligence

Jaeho Lee[*], Hong Guan[‡],
Michael Blumenstein[≠] and Yew-Chaye Loo[£]
Griffith School of Engineering, Griffith University

Abstract

Bridge Management Systems (BMSs) have been developed since the early 1990s, as a decision support system (DSS) for effective Maintenance, Repair and Rehabilitation (MR&R) activities in a large bridge network. Historical condition ratings obtained from biennial bridge inspections are major resources for predicting future bridge deteriorations through BMSs. However, available historical condition ratings are very limited in all bridge agencies. This constitutes the major barrier for achieving reliable future structural performances.

To alleviate this problem, a Backward Prediction Model (BPM) technique has been developed to help generate missing historical condition ratings. Its reliability has been verified using existing condition ratings obtained from the Maryland Department of Transportation, USA. This is achieved through establishing the correlation between known condition ratings and related non-bridge factors such as climate and environmental conditions, traffic volumes and population growth. Such correlations can then be used to determine the bridge condition ratings of the missing years. With the help of these generated datasets, the currently available bridge deterioration model can be utilized to more reliably forecast future bridge conditions.

In this chapter, the prediction accuracy based on four and nine BPM-generated historical condition ratings as input data are also compared, using traditional bridge deterioration modeling techniques, i.e., deterministic and stochastic methods. The comparison outcomes indicate that the prediction error decreases as more historical

---

[*] (j.lee@griffith.edu.au)
[‡] (h.guan@griffith.edu.au)
[≠] (m. blumenstein@griffith.edu.au)
[£] (y.loo@griffith.edu.au)

condition ratings are available. This implies that the BPM can be utilized to generate unavailable historical data, which is crucial for bridge deterioration models to achieve more accurate prediction results. Nevertheless, there are considerable limitations in the existing bridge deterioration models. Thus, further research is essential to improve the prediction accuracy of bridge deterioration models.

Keywords: Infrastructure Decision Support Systems (DSSs), Bridge Management Systems (BMSs), Bridge Condition Ratings, Deterioration Model, Maintenance, Repair and Rehabilitation (MR&R), Artificial Intelligence (AI), Artificial Neural Networks (ANNs), Backward Prediction Model (BPM)

INTRODUCTION

A bridge is usually designed to have long-term service life. In some cases, however, it could fail prematurely and as a consequence cause losses of human life. Thus, to ensure safety and optimal serviceability of a bridge, critical decision-making for Maintenance, Repair and Rehabilitation (MR&R) activities is required (Aktan et al., 2000). BMSs have been developed since the early 1990s as a Decision Support System (DSS) for effective MR&R activities in a large bridge network. Historical condition ratings obtained from biennial bridge inspections are major resources for predicting future bridge deteriorations through BMSs. However, available historical condition ratings are very limited in all bridge agencies. This constitutes the major barrier for determining reliable long-term structural performances. BMS generally assists significant future MR&R strategies, which are based on the results of a bridge deterioration model. Thus, an effective BMS highly relies on the prediction accuracy of deterioration rates (Madanat et al., 1995).

Many bridge condition ratings and deterioration models have been developed to determine the bridge lifecycle for major MR&R needs. Nevertheless, the predictions of future structural condition ratings from BMSs are still not practical for developing reliable long-term maintenance strategies. This is largely due to several drawbacks related to their application in most bridge agencies, viz: (1) commercial BMS software has been used for two decades, and bridge agencies would have roughly eight to nine biennial inspection records only; (2) bridge condition ratings usually do not change much during short-term periods; and (3) approximately 60% of BMS analytical process is affected by bridge inspection records (Lee et al., 2008a).

Ideally, a BMS should identify current and future bridge deficiencies and estimate the backlog of funding requirements. Typical BMS software mainly functions to (Hearn, 1999; Godart and Vassie, 1999): (1) forecast future bridge deficiencies; (2) identify a list of improvement options to correct such deficiencies; and (3) estimate the costs and benefits of implementing each improvement option. The condition ratings are used directly and indirectly as input data for many significant functions in the commercial BMS package (Hearn, 1999). Figure 1 presents the uses of bridge condition ratings and its relationship with many analytical BMS modules in project and network level analysis.

In relation to the lack of historical bridge condition ratings, most BMS software packages currently in operation worldwide suffer the following problems (Das, 1996; DeStefano and Grivas, 1998; Lounis and Mirza, 2001; Sianipar and Adams, 1997): (1) the predicted bridge condition ratings do not acceptably match the real situation; (2) interactive deterioration mechanism effects between or among structural elements are ignored; (3) one of the assumptions made in most deterioration models is that future conditions are dependent only on the current structural conditions because the past condition ratings are unavailable; (4) among BMS data requirements, the amount of time-dependent bridge datasets from periodic bridge inspections for updating a BMS is very limited; and (5) bridge condition rating variances in a small number of historical datasets inevitably lead to unreliable structural performances predicted by the current deterioration models in BMSs.

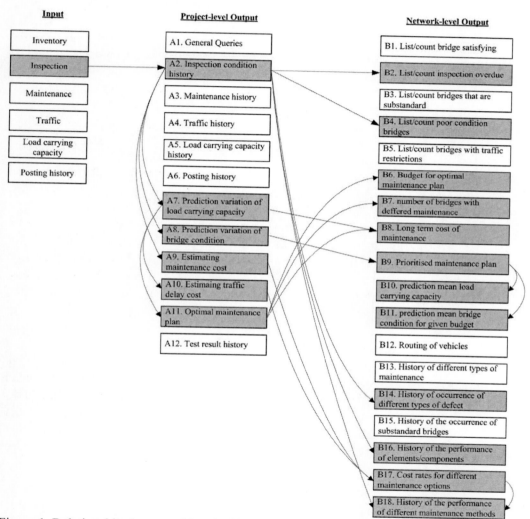

Figure 1. Relationships between historical bridge inspection data and BMS outputs (Godart and Vassie, 1999). (Note that the relationships with other input sources and BMS outputs are omitted).

In view of the above, current BMS outcomes can hardly be expected to be reliable. Many researchers and infrastructure asset management practitioners also have recognised that deterioration of infrastructure facilities is not deterministic (Mishalani and McCord, 2006).

These factors mainly lead to inaccuracy in predicting the future structural performance of bridges. Coupled with these drawbacks is the major weakness in current deterioration modelling techniques, which is essentially the lack of practical data related to the bridge element's modelling performance. These modelling techniques are invariably developed based on a few set of current structural condition ratings, thus unlikely to predict reliable future bridge condition ratings (Lee et al., 2008 a and b).

Time-Series Predictions for Insufficient Condition Rating Records

Time-series predictions are important resources for making decisions in many application domains (Weigend and Gershenfeld, 1994). Various prediction techniques have been applied to commercial BMS analysis modules. The most frequently used techniques are Regression, Markov models, Bayesian methods, Fuzzy techniques, Genetic Algorithms, Case-Based Reasoning and Artificial Neural Network (ANN) models. Specifically, Markov decision processes (MDP) have been used in major state-of-the-art BMS software as part of their deterioration modules. To obtain reliable predictions from conventional techniques, the size of missing patterns from an entire dataset must be 5% or less (Tabachinick and Fidell, 2001). It should be noted that irregularly sampled datasets cannot be used with conventional prediction methods (Karna et al., 2006). This research aims to utilize limited inspection records over a short period to predict large datasets spanning over a much longer time period. The short history of the BMS adoption and its lack of usable inspection records cause unreliable long-term bridge performance predictions. Recognizing the historical patterns for aging bridges can be a problem when using commonly available time series prediction methods.

For any computational prediction methods, the amount of available datasets is required to be much larger than the target prediction datasets to obtain reliable prediction results. An ANN-based model also requires a large number of training datasets to successfully estimate their correlation. It should be noted that existing ANN-based "data-mining" techniques have been applied in medical, economics, engineering and IT fields. While capable of carrying out similar activities as the BPM, data-mining has had success only in cases where very small proportions of the datasets are unavailable— much smaller than required to be generated for an effective BMS implementation. Because of the fundamental limitations in time-series predictions, the proposed neural network model adopts an alternative type of time-series dataset, which is used to overcome the lack of trends in the existing small number of bridge element condition ratings.

Outline of Research

To address the research problem identified, two steps of study have been proposed by the authors in an attempt to improve long-term predictions of the BMS. The first step involves generating a robust set of missing historical bridge condition ratings, which indicates the trend of structural condition depreciations. This is done by using the neural network-based Backward Prediction Model (BPM) based on the sample bridge data provided by the Maryland Department of Transport (DoT), USA. The BPM has an ability to produce missing historical condition ratings through the relationship between the actual condition ratings and non-bridge factors. In this respect, well-selected non-bridge factors are critical for the BPM to be able to obtain correct correlations. In the second step of the study, a reliable deterioration model will be developed based on complete historical condition ratings obtained from the results of the first step. The future bridge condition ratings predicted by this model will then be compared with the existing bridge data to determine the level of prediction accuracy. It should be noted that the BPM methodology was developed based on limited types of bridge elements. Although case studies have been conducted for verification, additional tests are required to further improve the BPM developed by using other types of bridge elements under different environment conditions.

METHODOLOGY OF GENERATING HISTORICAL CONDITION RATINGS

The research problem regarding the lack of historical bridge data can be solved by the use of an ANN-based Backward Prediction Model (BPM). A pilot study along these lines only considered bridge condition ratings amongst the BMS historical data required. The BPM predicts entire or selected periods of historical bridge condition ratings to generate unavailable years of datasets. It aims to improve the prediction accuracy of future bridge condition ratings using a deterioration module. The bridge condition ratings do not change much during a short period of time. As such, it is difficult to detect condition rating changes using an ANN-based condition rating prediction model. However, existing bridge condition ratings can be enhanced by non-bridge factors, including local climates, number of vehicles and population growth in the area surrounding the bridge. The non-bridge factors are employed to help establish the correlations between the non-bridge factors and the lack of historical data patterns in the existing but inadequate bridge condition rating datasets.

Figure 2 schematically describes the mechanism of the BPM. It illustrates the main function of the ANN technique in establishing the correlation between the existing condition rating datasets (from year m to year m+n) and the corresponding years' non-bridge factors. The non-bridge factors directly and indirectly affect the variation of the bridge conditions and thereby the deterioration rate. The relationships established using neural networks are then applied to the non-bridge factors (for year zero to year m) to generate the missing bridge condition ratings (for the same year zero to year m). Thus, the non-bridge factors in conjunction with the ANN technique can produce the historical trends that inform the current condition ratings (Lee et al., 2008a).

The training algorithm used in the BPM is the back-propagation algorithm, and it is used in conjunction with a feed-forward multi-layer neural network. It is a universal-approximator function that typically yields better results than traditional approximation methods in practical applications. It is also most commonly used in solving non-linear engineering problems. The log-sigmoid function, as a typical neuronal non-linear transfer function, is used in the proposed model because of its non-linearity.

As shown in Figure 3, the input layer has such non-bridge factors as the number of vehicles, population growth and climatic conditions; it incorporates 21 variables including four factors for the traffic volume, two factors for the population growth and 15 factors for the climates. This information is used to train the ANN to determine the correlation with currently available bridge condition rating data in the output layer. The specifications for the inputs, outputs and functions of the BPM are also detailed in Table 1.

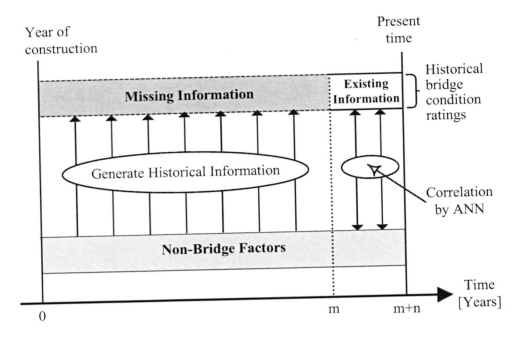

Figure 2. The mechanism of the BPM

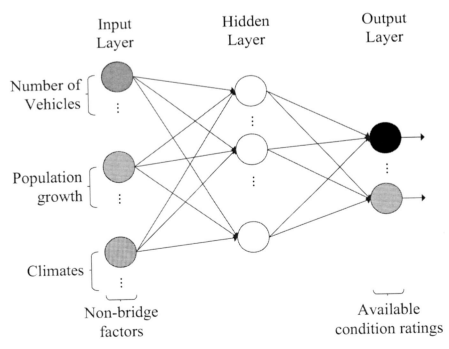

Figure 3. Structure of ANN-based BPM

Table 1. Components of the Proposed Neural-Network Model

Training Algorithm	Back Propagation Algorithm
Transfer Function	Log-sigmoid Function
Inputs	Traffic volume (four factors), Population growth (two factors) and Climates (15 factors)
Total number of neurons	Input: 21 @ each year, Output: one @ each year
Hidden layers	two (15 neurons in first layer and four neurons in second layer)
Output	Bridge Condition Ratings (one output @ each year)
Scale of learning rate (lr) and Momentum coefficient (mc)	0.0 - 0.5 in 0.1 steps (six cases) / 0.0 - 1.0 in 0.1 steps (11 cases)
Total number of cases generated	66 cases (combination of lr and mc) @ each year

Sample Condition Ratings for BPM

The BPM has been tested using two different types of bridge condition rating datasets - the National Bridge Inventory (NBI) and BMS condition rating inputs— for the same bridge provided by the Maryland Department of Transport (DoT), USA. The obtained bridge condition datasets require calibration to fit into the BPM model, due to the typical ANN input environment. The acceptable numerical scale for ANN modelling is from -1 to 1 (or 0 to 1). Figure 4 illustrates the scale of NBI and element level-based condition rating information for this particular study. The Condition Index (CI) in NBI is scaled between zero to nine for the NBI#58 (deck), #59 (superstructure), and #60 (substructure), and every calibration step is titled as a different Condition State (CS) to express the bridge component's condition ratings. The CI for an element level inspection method consists of four or five different CSs (depending on the bridge authorities' adoption and customisation) to quantify condition states of bridge elements (Lee et al., 2008a).

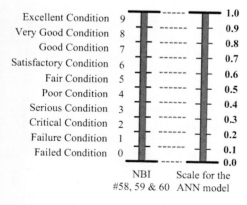

(a) Scales of NBI

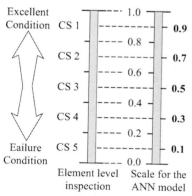

(b) Scales of BMS element condition rating inputs

Figure 4. Scales of tested condition ratings for the BPM (Lee et al., 2008a)

Generating Historical Condition Ratings Using National Bridge Inventory Records

Two validation methods have also been established: backward-manner comparison (Test #1); and forward-manner comparison (Test #2). The backward-manner comparison is the BPM outcome that can be directly compared with existing historical data to measure its prediction accuracy. This comparison is used with the NBI when the BPM methodology is established. The forward-manner comparison is the generated condition rating data, which feeds into the BPM again as input to predict present year's condition ratings. The outcome will be compared with the existing data, and thereby the generated past year's condition ratings can also be verified.

The entire timeframe of the bridge data used in the BPM is from year 1966 to 2004, as shown in Figure 5. The figure shows the timeframe for backward comparisons (Test #1) and forward comparisons (Test #2) for the BPM using NBI information. This includes the timeframe of inputs (Figure 5 (c) and (f)), outputs (Figures 5(d) and (g)) of backward comparisons (Test #1, Figure 5 (e)) and forward comparisons (Test #2, Figure 5(h)).

Amongst these, on five occasions, inspection results have been used as BPM training inputs and outputs (from 1996, in two-year increments to 2004). The assumed condition rating (excellent condition) at year zero (1966) of the bridge has also been used. The remaining years (1968 to 1994, with two-year increments) of historical condition ratings are generated. The results of a BPM (1968 to 1994) have showed reliable back-prediction performance. The generated historical condition ratings are compared with the existing NBI datasets (1968 to 1994: 14 missing inspection records) to measure its prediction errors. The BPM generated 73.7% of the missing years' condition ratings using 26.3% of the available bridge condition ratings. That is: the missing/entire inspection records = 14/19; the existing/entire inspection records = 5/19. Two tests (tests #1 and 2) are conducted to validate this model, and associated discussions are detailed in the next two sub-sections.

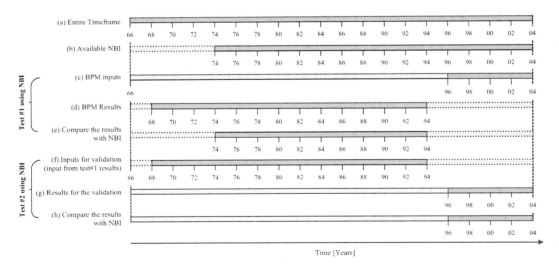

Figure 5. BPM timeframe (Test #1 and 2 using NBI for performance measurements)

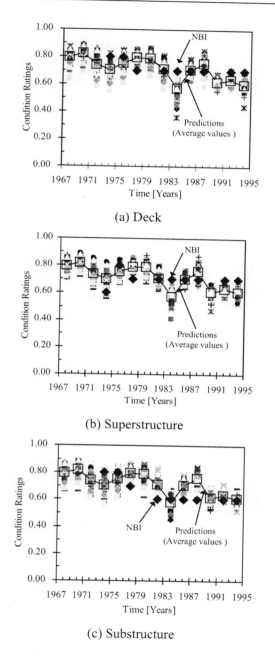

(a) Deck

(b) Superstructure

(c) Substructure

Figure 6. BPM results for Bridge#0312xxx1. (Note that the number of prediction results in each year is 66, which is the combined number of learning rates (lr: 0.0-0.5) and momentum coefficients (mc: 0.0-1.0) in the neural network configuration)

Backward Comparisons (Test #1)

As mentioned above, most of the generated data from Test #1 can be directly compared with the existing historical NBI datasets. The comparisons between the results for each bridge component and its NBI records are plotted in Figure 6. Most artificially-generated historical condition ratings are obtained within a prediction error scale of less than ±10%. However, year 1982, for deck, year 1984, for superstructure, and years 1982, 1984 and 1986, for substructure exhibit larger errors than the maximum allowance ($\pm 10\%$). The main reason for generating inaccurate outcomes is that the proposed model is developed based on the correlations of condition ratings and their corresponding non-bridge factors in the ANN training stage. Nevertheless, this is considered adequate for historical condition ratings, because they are ranked within the same Condition State ($60\% \leq CS2 < 80\%$).

In the case where the ANN training datasets, using existing condition ratings, do not have a relevant correlation with the non-bridge factors, the BPM cannot provide reliable historical condition ratings for a specific year. For example, depreciations of condition ratings caused by sudden physical damages to a bridge are not related to the non-bridge factors used in the BPM. Hence in such circumstance, the BPM is unable to generate accurate historical condition ratings. The results of the backward predictions are validated by comparing them with existing historical condition ratings (Test #1). However, the actual element-level condition ratings for BMS inputs are only available in a small number of datasets and are not applicable to the backward comparison method. As such, the forward comparison (Test #2) is used to validate the BPM.

Table 2. Performance measurements: existing NBI vs BPM results (Bridge # 0312xxx1)

Year	Deck			Superstructure			Substructure		
	A	B	C	A	B	C	A	B	C
1968	0.793			0.793			0.794		
1970	0.822			0.822			0.822		
1972	0.741			0.737			0.739		
1974	0.704	0.8	0.0955	0.699	0.6	0.0995	0.703	0.8	0.0966
1976	0.754	0.8	0.0464	0.753	0.8	0.0472	0.753	0.8	0.0471
1978	0.796	0.7	0.0961	0.792	0.7	0.0923	0.797	0.7	0.0966
1980	0.787			0.788			0.788		
1982	0.711	0.7	0.0108	0.708	0.7	0.0076	0.714	0.6	0.1137
1984	0.571	0.7	0.1291	0.574	0.7	0.1259	0.579	0.6	0.0208
1986	0.703	0.7	0.0034	0.701	0.7	0.0013	0.704	0.6	0.1043
1988	0.755	0.7	0.055	0.759	0.7	0.0585	0.757	0.6	0.1573
1990	0.617			0.611			0.616		
1992	0.634	0.7	0.0661	0.63	0.7	0.0695	0.635	0.6	0.0347
1994	0.601	0.7	0.0988	0.607	0.7	0.0933	0.606	0.6	0.0059

Note - A: Average predictions, B: NBI, C: Difference between A and B

Forward Comparisons (Test #2)

The BPM of the ANN training inputs in Test #2 utilizes the results of Test #1 between 1968 and 1994, as illustrated in Figure 5 (f). The BPM produces the future condition ratings between 1996 and 2004. The results are also compared with the existing condition ratings (year 1996-2004) which are summarized in Table 3. The condition ratings, predicted by using the ANN-based BPM, provide satisfactory results within the error allowances (i.e., $\pm 10\%$). Therefore, the forward comparison method can be used to validate the BPM results using the actual BMS condition rating inputs.

Table 3. Summary of prediction performance for forward comparisions

Year	Deck A	B	C	Superstructure A	B	C	Substructure A	B	C
1996	0.568	0.6	0.032	0.567	0.6	0.033	0.569	0.6	0.031
1998	0.59	0.6	0.01	0.593	0.6	0.007	0.591	0.6	0.009
2000	0.665	0.6	0.065	0.666	0.6	0.066	0.669	0.6	0.069
2002	0.59	0.6	0.01	0.595	0.6	0.005	0.591	0.6	0.009
2004	0.555	0.6	0.045	0.556	0.6	0.044	0.557	0.6	0.043
Mean Errors (%)	-	-	3.2	-	-	3.1	-	-	3.2

Note - A: Average predictions, B: NBI, C: Difference between A and B

Generating Historical Condition Ratings Using BMS Condition Rating Records

The actual BMS inputs only have a small number of datasets and are not applicable to the backward-manner comparison. Hence, the forward-manner comparison is the only alternative to determine the accuracy of the BPM outcomes when using the limited BMS condition ratings as input value to the BPM. To exemplify this, the BPM test for BMS actual inputs is conducted using one superstructure element of the BMS condition rating datasets—Element #234: Reinforced Concrete Pier Cap. This data has been collected for the Maryland DOT's periodically updated BMS database. Figure 7 describes the BPM timeframe for the bridge Element #234, which shows the time (in the number of years) for: (a) the entire bridge life cycle; (b) available condition ratings; (c) BPM inputs; (d) generated historical condition ratings; (e) inputs for validation; (f) forward-prediction results; and (g) results comparison of the forward-predictions with the existing condition rating datasets.

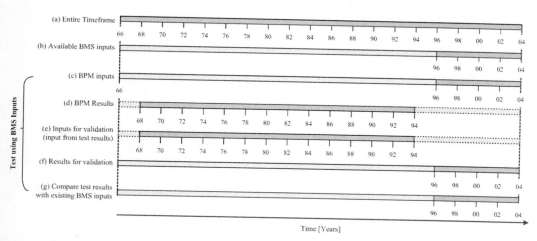

Figure 7. BPM timeframe of Element #234 on Bridge #3210xxx1

Only five historical condition rating datasets (from 1996 to 2004, with two-year increments) are available as BPM input values as detailed in Table 4. The Condition Index (CI) of BMS condition ratings is scaled between 0% (fail) and 100% (excellent) as shown in

Figure 4. The data requires calibration to fit into the BPM for the typical ANN input environment. The acceptable numerical scale for ANN modelling is from −1 to 1, as shown in Figure 4.

Table 4. Raw data of actual BMS inputs (Element #234 on Bridge #3210XXX1)

Year of inspection	Total quantity (%)	CS1 (%)	CS2 (%)	CS3 (%)	CS4 (%)	CS5 (%)
1996	100	80	14	6	0	0
1998	100	80	14	6	0	0
2000	100	80	14	6	0	0
2002	100	80	19	1	0	0
2004	100	80	19	1	0	0
Average	100	80	16.2	3.8	0	0

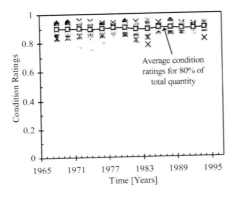

(a) 80% of the total quantity

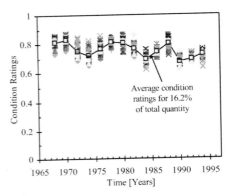

(b) 16.2% of the total quantity

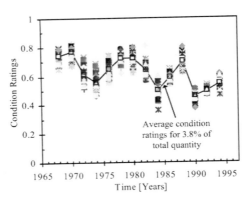

(c) 3.8% of the total quantity

Figure 8. Backward prediction results for Element #234 for Bridge #0312xxx1. (Note that in the figure, there are 66 prediction results in each year, being derived from the combined number of learning rates (lr: 0.0-0.5) and momentum coefficient (mc: 0.0-1.0) in the neural network.)

Table 4 shows that the average quantity of each Condition State (CS) on Element #234 between 1996 and 2004, is about 80%, 16.2% and 3.8% of the total elements for CS1, CS2 and CS3, respectively. The BPM generates historical condition ratings from 1968 to 1994, in three different proportions of the condition state, as shown in Figure 8. To validate the results of the BPM (1968-1994), the forward-manner comparison (Test #2) is used based on the NBI datasets. This is because the condition ratings between 1967 and 1994, are not available for direct comparison.

For the forward-manner comparison (Test #2), the back-prediction results (1968 to 1994) are used as input datasets to generate the condition ratings for the subsequent years 1996 to 2004. The BPM-generated condition ratings are then directly compared with the existing condition rating datasets—the details of raw data for BMS inputs are shown in Table 4. The BPM results are shown in Figure 9 for four different proportions of the element quantity. Table 5 summarizes the final results from the BPM as well as its prediction errors for years 1996 to 2004. The yearly average prediction errors are less than ±10%. It seems that forward comparison method (Test #2) has confirmed the reliability of the BPM in generating historical BMS condition ratings for years 1968 to 1994.

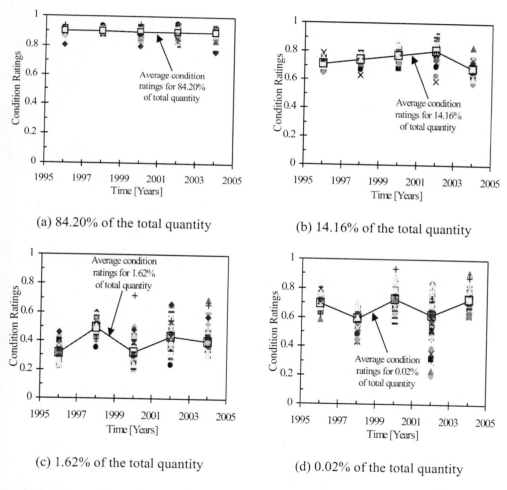

(a) 84.20% of the total quantity   (b) 14.16% of the total quantity

(c) 1.62% of the total quantity   (d) 0.02% of the total quantity

Figure 9. Performance measurements of Element #234 for Bridge #0312xxx1. (Note that in the figure, there are 66 prediction results in each year, being derived from the combined number of learning rates (lr: 0.0-0.5) and momentum coefficient (mc: 0.0-1.0) in the neural network.)

Table 5. Prediction errors of the BPM using forward comparisons

Year		CS1 (%)	CS2 (%)	CS3 (%)	CS4 (%)	CS5 (%)	Total (%)
1996	A	84.2	14.18	0.1	1.52	0	100
	B	80	14.29	5.71	0	0	100
	C	4.2	0.11	5.62	1.52	0	
	D	2.29					
1998	A	84.41	13.98	1.58	0.02	0	100
	B	80	14.29	5.71	0	0	100
	C	4.41	0.31	4.13	0.02	0	
	D	1.78					
2000	A	87.65	10.76	0.2	1.4	0	100
	B	80	14.29	5.71	0	0	100
	C	7.65	3.53	5.52	1.4	0	
	D	3.62					
2002	A	91.29	6.96	1.2	0.54	0	100
	B	80	19.14	0.86	0	0	100
	C	11.29	12.18	0.34	0.54	0	
	D	4.87					
2004	A	81.86	16.35	0.85	0.93	0	100
	B	80	19.14	0.86	0	0	100
	C	1.86	2.79	0	0.93	0	
	D	1.12					

Note - A: Predictions, B: existing BMS condition ratings, C: Difference between A and B, D: average difference

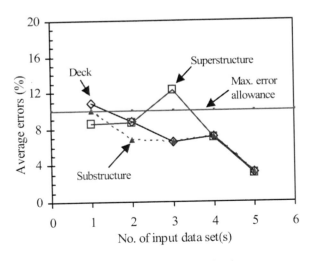

Figure 10. Average errors for different numbers of ANN training inputs

Minimum Data Requirement for BPM

To verify the minimum number of the neural network training input requirements and the response with the number of inspection records in the BPM, additional tests are also conducted using the NBI. Figure 10 shows the forward comparisons when the BPM is used for one to five sets of condition ratings as input. The figure demonstrates the average prediction errors, which gradually decrease as the number of inspections increases. The effective range of inputs for the number of inspections starts from two sets of records, which occurs prediction errors below the maximum error allowance (i.e., ±10%). It is demonstrated that the proposed BPM can provide satisfactory results when more than four sets of inspection records are used as its input values.

Summary of the Case Studies

Further verification of the BPM has been also carried out via nine case studies, i.e., nine typical bridge elements from seven bridges owned by the Roads and Transport Authority of New South Wales (RTA NSW) under three different condition state scales, i.e., CS 3, 4 and 5. The historical bridge inspection records are provided by the RTA. The corresponding years of non-bridge factors, including climates and population growth for these case studies, are provided by the Australian Government Bureau of Meteorology and the Australian Bureau of Statistics (ABS). Sample bridges data obtained were built during the 1960s and 1970s, with an approximate average life cycle of 40 years. The bridge element types in these case studies are defined in the RTA bridge inspection procedure manual (RTA NSW, 1999). The average number of bridge element-level inspection datasets obtained for the BPM is between four and six records for the 10 to 12 years of historical bridge element condition ratings.

The nine typical bridge elements are used to demonstrate the capability of the BPM. The test methods of generating the missing bridge elements' condition ratings and their validations are identical to those presented in the previous sections. As shown in Table 6, the maximum prediction errors do not exceed the pre-defined maximum allowable limit for the nine bridge elements. The maximum average yearly prediction errors are found to be 10.21%, 9.26% and 4.40%, respectively, for the three different condition states (i.e., CSs1, 2 and 3). These BPM results are considered satisfactory.

Table 6. Average condition rating differences of the nine typical bridge elements

Bridge element	No. of condition states (CSs)	Max. error allowance (%)	Prediction error (%)		
			Min.	Max	Mean
Assembly Joint Seal	3	33.33	1.69	2.50	
Brick / Masonry / Reinforced Earth	3	33.33	3.95	18.61	10.21
Elastomeric Bearing Pad	3	33.33	1.14	9.52	
Concrete-Deck Slab	4	25.00	0.71	4.18	
Concrete-Pile	4	25.00	11.03	12.21	
Concrete-Pier	4	25.00	1.63	4.47	9.26
Concrete-Pre-tensioned Girder	4 .	25.00	0.73	1.56	
Metal Railing	4	25.00	5.81	23.89	
Steel(L)-Beam / Girder	5	20.00	1.14	4.40	4.40

Typical Bridge Deterioration Models

Many research studies on bridge deterioration models have been carried out to improve the reliability of BMS outcome. Nonetheless, the successful achievement of the analysis using these models remains highly dependent on the quality and sufficiency of data gathered (Kleywegt and Sinha, 1994).

According to Morcous et al. (2002), current bridge deterioration models can be categorized as deterministic, stochastic and artificial intelligence. In this section, only the first two modeling techniques are considered, as they are most common in many current BMSs. A summary of these two modelling techniques are given in Table 7. Generally, a deterministic model predicts that a bridge will deteriorate with regard to a particular algorithm, while a stochastic model considers that actual deterioration rate is unknown and contains a probability that the bridge will deteriorate at a particular rate (Austroads, 2002). Among the deterministic models, regression analysis is a methodology widely used in many BMSs (Kleywegt and Sinha, 1994), whereas Markovian-based model is considered as the most common method among the stochastic techniques (Micevski et al., 2002). Therefore, these two models are used in the current study to predict future bridge conditions based on the BPM-generated historical data (Son et al., 2009).

Table 7. Categories of Bridge-Deterioration Models (Morcous et al., 2002)

Categories	Methodology	Details
Deterministic	Straight-line	-
	Extrapolation	Stepwise regression
	Regression	Linear regression
		Nonlinear regression
	Curve-fitting	B-spline approximation
		Constrained least squares
Stochastic	Simulation	-
	Markovian	Percentage prediction
		Expected-value method
		Poisson distribution
		Negative-binomial model
		Ordered-probit model
		Random-effects model
		Latent Markov-decision process

Comparison of Models

In this section, the evaluation of prediction accuracy obtained from both linear and non-linear regression analyses, as well as Markovian-based model are presented. Generally, the determination of a functional form of the equation that could fit particular datasets (also referred to as a performance curve) is considered as crucial part of regression modeling (Austroads., 2002). As for linear regression, this function is expressed by a simple linear equation; whereas in non-linear regression, this function is characterized as a polynomial form of second or more orders. In this study, following Jiang and Sinha (1989), only a third-order polynomial model is used to determine long-term deterioration of bridge condition ratings. Eqn. 1 presents a performance curve of bridge element using a third-order polynomial.

$$C(t) = \beta_0 + \beta_1 t_i + \beta_2 t_i^2 + \beta_3 t_i^3 + \alpha_i \tag{1}$$

where, C(t) = condition rating of a bridge at age t; t_i = bridge age ; α_i = error term; and β_0 = recorded condition rating of a new bridge.

The predictions from both linear and non-linear regressions are also carried out using four generated BMS datasets by BPM (from years 1978 to 1984), as shown in Figure 11.

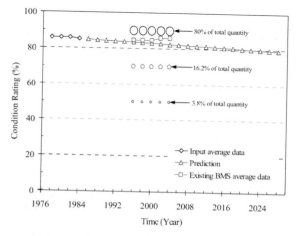

(a) Linear regression: 3.5% average errors

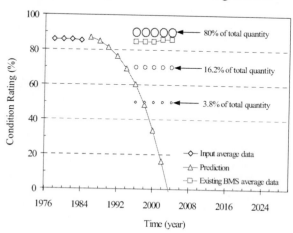

(b) Non-linear regression: 74.4% average errors

Figure 11. Prediction results using four of BPM-generated condition ratings (from 1978 to 1984)

The average prediction error of linear regression is obtained by averaging the differences between the condition ratings of the existing BMS condition ratings and the prediction data from 1996 to 2004. Similar method is employed to calculate the average prediction error of non-linear regression. As a result, the average prediction errors of linear and non-linear regression models are 3.5% and 74.4%, respectively. It should, however, be noted that the prediction results generated by non-linear regression technique show unusual pattern of deterioration, as illustrated in Figure 11(b). This might be resulted from the very limited number of input data used in the prediction.

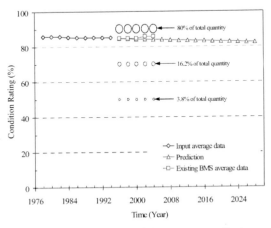

(a) Linear regression: 1.5% average prediction errors

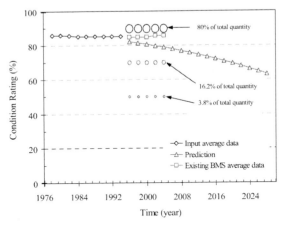

(b) Non-linear regression:4.7% average prediction errors

Figure 12. Prediction results using nine sets of BPM-generated condition ratings (from 1978 to 1994)

Figure 12 presents the prediction results based on nine historical data records generated by the BPM (from 1978 to 1994). It should be noted that the BPM-based historical condition ratings are generated as 66 combinations of learning rate and momentum coefficient. In order for these results to be used in the regression analysis, the 66 combinations in each of the year 1968 to 1994, are averaged to represent each type of bridge elements' individual condition rating records. Following this, the existing BMS condition ratings and the BPM-based prediction results are compared to evaluate the prediction accuracy. Following the similar approach mentioned above, the average prediction errors between the generated condition ratings and the BMS condition ratings are calculated for both linear and non-linear regression models. This yielded the average prediction errors are 1.5% and 4.7%, respectively. The predictions are much improved as compared to using four sets of historical condition ratings.

Figure 13 shows the prediction results of the Markovian-based model based on four historical data records generated by the BPM. Theoretically, the Markovian-based model predicts bridge condition ratings using the probabilities of bridge conditions transition. These probabilities are characterised in a matrix type, namely, the transition probability matrix. If the current state of bridge conditions or the initial state is known, condition from one rating to another can be forecasted throughout multiplication of original state vector and the transition matrix (Jiang and Sinha, 1989). To estimate the transition probabilities, the subsequent nonlinear programming objective function is formulated (Jiang and Sinha, 1989):

$$\min \sum_{t=1}^{N} |A(t) - E(t,P)| \tag{2}$$

subject to $0 \leq p(i) \leq 1$ i= 1,2,...,U

where, N = 6, the number of years in one bridge age group; U = 6, the number of unknown probabilities; P = a vector of length I equal to [p(1), p(2), ...,p(I)]; A(t) = the average of condition ratings at time t, estimated by regression function; and E(t,P) = estimated value of condition rating by Markov chain at time t.

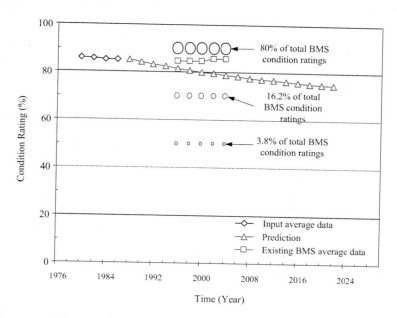

Figure 13. Prediction results using four BPM-generated historical data in Markovian-based model

Table 8 compares the errors of the predictions using four and nine BPM-generated condition ratings, for linear regression, non-linear regression and Markovian-based models. It is evident that, for all modelling techniques, the prediction errors decrease as more input data become available. In the case of linear regression, the average error of 3.5% from the prediction using four BMS condition ratings decreases to 1.5% when nine generated condition ratings are used. Similarly, for the case of non-linear regression, the prediction error decreases from 74.4% to 4.7% when the number of input datasets increases. As for Markovian-based model, only four sets can be used as input data for predictions. The average prediction errors of Markovian-based model are 7.0%.

Table 8. Comparison Of Prediction Errors Using Four and Nine Bpm-generated Data

Prediction techniques	Number of input data	Difference between prediction and existing BMS condition ratings
Linear regression	A	3.5%
	B	1.5%
Non-linear regression	A	74.4%
	B	4.7%
Markovian-based model	A	7.0%
	B	NA

A: four-generated data used in BPM; B: nine-generated data used in BPM

The above findings indicate that the amount of input datasets is essential for typical numerical prediction methods to gain dependable prediction results. The findings also suggest that, in both deterministic and stochastic models, the historical data generated by the BPM technique can contribute to the improvement of prediction accuracy. This reinforces the applicability of the BPM in generating missing historical condition ratings that are capable of providing a basis for more reliable predictions of future bridge conditions.

Notwithstanding the above findings, several limitations of the above models are also worth noting. As for the deterministic models, these are: (1) the models disregard the uncertainty due to the stochastic nature of bridge deteriorations (Jiang and Sinha, 1989); (2) they predict the average condition of a bridge structure rather than the current and historical condition ratings of individual elements; (3) they approximate bridge structure deterioration only for the case of "no maintenance" strategy because it is difficult to estimate the influence from various maintenance strategies (Sanders and Zhang, 1994); (4) they ignore the interaction between the different bridge structure elements, for example, between the bridge deck and the deck joints (Sianipar and Adams, 1997); and (5) they are difficult to be revised when new condition ratings are gained (Morcous et al., 2002).

Although the Markovian-based models can address two problems in deterministic models by capturing the uncertainty of the deterioration process and accounting for the current facility condition in predicting the future one, they still suffer from the following limitations: (1) Markovian-based models currently implemented in advanced BMS use the first-order Markovian Decision Process that assumes state independence for simplicity (DeStefano and Grivas, 1998), which means that the future facility condition depends only on the current facility condition and not on the facility condition history, which is unrealistic (Madanat et al., 1997); (2) transition probabilities assume that the condition of a facility can either stay the same or decline, thus avoiding the difficulty of estimating transition probabilities for facilities where treatment actions are performed (Madanat and Ibrahim, 1995); (4) Markovian-based models cannot efficiently consider the interactive effects between the deterioration mechanisms of different bridge components (Sianipar and Adams, 1997); and (5) transition probabilities require updates when new data are obtained as bridges are inspected, maintained, or rehabilitated, which is a time-consuming task (Morcous et al., 2002).

CONCLUSION

The development of the ANN-based Backward Prediction Model (BPM) is described in some detail in this chapter. The BPM uses non-bridge factors as supplementary historical data to overcome the lack of historical bridge data in terms of its quantity and patterns. The non-bridge factors, including local climates, traffic volume and population growth in the area surrounding the bridge, are employed to help establish the correlations between the non-bridge factors and the lack of historical data patterns in the existing condition ratings. The results obtained from the BPM have been validated by using both backward and forward comparison techniques. The former compares the BPM outcomes with the known historical data to assess the prediction accuracy, whereas the latter uses the BPM outcomes as input data to predict present year's bridge condition ratings, which are then compared directly with the actual data in such year. To carry out the backward comparison, sample structural condition rating datasets made available by the Maryland Department of Transportation (Maryland DoT), and based on five existing condition rating datasets (years 1996 to 2004 and 26% of the total record), ensured that the BPM is able to generate 14 missing datasets (years 1968 to 1994 and 74% of the total records) for the intervening years when proper inspection records are missing. The average ratio of the generated and existing datasets is about three. The average prediction errors of the generated bridge condition ratings are between 6.7% to 7.5% over a period of 20 years.

To confirm the BPM methodology established, case studies are also performed using nine typical bridge elements from the Road and Traffic Authority of New South Wales (RTA NSW). Same validation methods, i.e., backward and forward comparisons, are used to generate missing bridge elements' condition ratings. The maximum yearly prediction errors of three different condition state scales (CS 3, 4 and 5) are 18.61%, 23.89% and 4.40%, respectively. These are satisfactory as compared to the maximum errors of 33.33%, 25% and 20% for 3CSs, 4CSs and 5CSs, respectively.

In addition, the methodology described in this chapter aims to establish a possible solution and a very initial approach for determining the unavailable historical structural condition ratings. The performance of BMSs for optimal MR&R strategy relies heavily on bridge deterioration models, which in turn depends on the quality and sufficiency of data gathered. The lack of historical bridge condition ratings is a major problem encountered by the current deterioration modelling to achieve reliable prediction of future bridge conditions. To overcome this drawback, the Backward Prediction Model (BPM) is introduced in this chapter as a means to assist in generating unavailable historical condition data, which is achieved by correlating existing bridge condition dataset with non-bridge factors. Consequently, further study may be required to explore the types and numbers of non-bridge factors for the BPM. As further critical non-bridge factors are identified and incorporated into the model, the accuracy of correlations with a small number of structural condition ratings, as well as future condition rating predictions, will improve.

The BPM outcomes are also applied to most typical bridge deterioration models to find advantages of using BPM. To ensure that the quality of such generated data is sufficient, future prediction results using the generated data is compared with those five existing BMS condition ratings. Under both linear and non-linear regression deterioration modelling scenarios, the average errors of the prediction results using nine BPM-generated historical condition records are less than those using four BPM-generated records. This indicates that the prediction errors become smaller as the amount of input data increases. Hence, using BPM to generate more historical condition data could contribute to improved prediction of future bridge conditions. These findings, however, should be interpreted in light of the following main limitations of the deterministic deterioration models employed in this study: (1) their prediction is based only on an average condition of a bridge structure with no regard to the variability of condition rating distributions in each year; and (2) they disregard the interaction between the different bridge structure elements. As for the Markovian-based model, prediction using four BPM-generated historical condition records is compatible with prediction of regression deterioration model using nine BPM-generated records. However, prediction of Markovian-based models is typically unrealistic due to ignorance of historical condition ratings. Moreover, the proportion of each condition state is not considered in prediction. This means the risk of individual bridge elements is also ignored. Thus, further research is essential to deal with such limitations and develop a more robust deterioration model that fully makes use of the benefits of the BPM-generated historical condition records.

REFERENCES

Aktan, A. E., Catbas, F. N., Grimmelsman, K. A., and Tsikos, C. J. (2000). "Issue in infrastructure health monitoring for management." *Journal of Engineering Mechanics*, 126(7), pp711-724.

Austroads. (2002). *"Bridge Management Systems – The State of the Art."* Austroads: AP-R198/02, pp23.

Das, P.C. (1996). *"Bridge Maintenance Management Objectives and Methodologies.* Bridge Management 3: inspection, maintenance, assessment and repair, E&FN Spon: London, pp1-7.

DeStefano, P. D., and Grivas, D. A. (1998). "Method for estimating transition probability in bridge deterioration models." *Journal of Infrastructure Systems*, 4, pp56-62.

Godart, B. and Vassie, P.R. (1999). *"Review of existing BMS and definition of inputs for the proposed BMS."* Deliverable D4: BRIME Report.

Hearn, G. (1999). *"Segmental Inspection for Improved Condition Reporting in BMS."* In the Proce. of the Eighth Int. Bridge Management Conference, Denver, Colorado, B-3/1-8.

Jiang, Y., and Sinha, K. C. (1989). *"Bridge service life prediction model using the Markov Chain"* Transportation Research Record1223, pp24-30

Karna, T., Rossi, F. and Lendasse, A. (2006). *"LS-SVM functional network for time series prediction."* In the Proce. of the of XIVth European Symposium on Artificial Neural Networks (ESANN 2006), Bruges. Belgium, pp473-478.

Kleywegt, A. J., and Sinha, K. C. (1994). *"Tools for bridge management data analysis."* Transportation Research Circular (423), pp16-26.

Lee, J.H., Sanmugarasa, K., Blumenstein, M. and Loo, Y. C. (2008a). "Improving the Reliability of a Bridge Management System (BMS) using an ANN-based Backward Prediction Model (BPM)." *Journal of Automation in Construction*, 17(6), pp758-772.

Lee, J. H., Guan, H., Blumenstein, M., and Loo, Y. C. (September, 2008b*). "An ANN-Based Backward Prediction Model for Reliable Bridge Management System Implementations Using Limited Inspection Records-Case Studies."* 17th Congress of IABSE: Creating and Renewing Urban Structures-Tall Buildings, Bridges and Infrastructure, International Association for Bridge and Structural Engineering (IABSE), Chicago, USA, CD-ROM Proceeding (Paper#.A-1144).

Lounis, Z. and Mirza, M. S. (2001*). "Reliability-based service life prediction of deteriorating concrete structures."* In the Proce. of 3rd Int. Conference on Concrete Under Severe Conditions, University of British Columbia, Canada.

Madanat, S. M., Karlaftis, M. G., and McCarthy, P. S. (1997). "Probabilistic infrastructure deterioration models with panel data." *Journal of Infrastructure Systems*, 3, pp4-9.

Madanat, S., and Ibrahim, W. H. W. (1995). "Poisson regression model of infrastructure transition probabilities." *Journal of Transportation Engineering*, 121, pp267-272.

Madanat, S., Mishalani, R., and Ibrahim, W. H. W. (1995). "Estimation of Infrastructure Transition Probabilities from Condition Rating Data." *Journal of Infrastructure Systems*, 1, pp120-125.

Micevski, T., Kuczera, G., and Coombes, P. (2002). "Markov Model for Storm Water Pipe Deterioration." *Jounal of Infrastructure System*, 8, pp49-56.

Mishalani, R. G., and McCord, M. R. (2006). "Infrastructure Condition Assessment, Deterioration Modeling, and Maintenance Decision Making: Methodological Advances and Practical Considerations." *Journal of Infrastructure Systems*, 12(3), pp145-146.

Morcous, G., Rivard, H., and Hanna, A. M. (2002). "Modelling Bridge Deterioration Using Case-based Reasoning." *Journal of Infrastructure Systems*, pp86-95.

RTA NSW. (1999). *"RTA Bridge Inspection Procedure."* Roads and Transport Authority of New South Wales, Australia.

Sanders, D. H., and Zhang, Y. J. (1994). *"Bridge deterioration models for states with small bridge inventories."* Transportation Research Record 1442, pp101-109.

Sianipar, P. R. M., and Adams, T. M. (1997). "Fault-Tree Model of bridge element deterioration due to interaction." *Journal of Infrastructure Systems*, 3, pp104-110.

Son, J.B., Lee, J. H., Blumenstein, M., Loo, Y.C., Guan, H. and Panuwatwanich, K., (2009). *"Improving reliability of Bridge deterioration model using generated missing condition ratings."* 3rd International Conference on Construction Engineering and Management (ICCEM), Jeju, S. Korea, 27-30 May, 2009, CD-ROM Proceeding, Paper#S13-1

Tabachinick, B.G. and Fidell, L.S. (2001). *"LS-SVM functional network for time series prediction."* In the Proce. of the of XIVth European Symposium on Artificial Neural Networks (ESANN 2006), Bruges. Belgium, pp473-478.

Weigend, A. and Gershenfeld, N. (1994). *"Time Series Prediction: Forecasting the Future and Understanding the Past,"* Addison-Wesley, Reading, MA, pp59-66.

INDEX